A Text Book Of

SOLID WASTE MANAGEMENT

Course Code 22605

Semester - VI

THIRD YEAR DIPLOMA IN CIVIL ENGINEERING GROUP

As Per MSBTE's 'I' Scheme Syllabus

Dr. D. K. Gupta
Ph.D., PE, MASCE
Government Polytechnic
Washim

V. K. Sonarkar
B.E. (Civil), M.E. (Envi. Engg.)
V.B.V. Polytechnic,
Vasai Road (W), Dist. Palghar.

S. B. Nimbalkar
B.E. (Civil), M.E. (Envi. Engg.)
Ex. Lecturer,
Government Polytecyhnic,

NIRALI PRAKASHAN
ADVANCEMENT OF KNOWLEDGE

N4553

Solid Waste Management **ISBN 978-93-89686-83-8**

First Edition : January 2020

Published By :

NIRALI PRAKASHAN

Abhyudaya Pragati, 1312, Shivaji Nagar

Off J.M. Road, PUNE – 411005

Tel - (020) 25512336/37/39, Fax - (020) 25511379

Email : niralipune@pragationline.com

PUNE

Nirali Prakashan : 119, Budhwar Peth, Jogeshwari Mandir Lane, Pune 411002, Maharashtra
Tel : (020) 2445 2044
Email : bookorder@pragationline.com, niralilocal@pragationline.com

Nirali Prakashan : S. No. 28/27, Dhyari, Near Pari Company, Pune 411041
Tel : (020) 24690204 Fax : (020) 24690316
Email : dhyari@pragationline.com, bookorder@pragationline.com

MUMBAI

Nirali Prakashan : 385, S.V.P. Road, Rasdhara Co-op. Hsg. Society Ltd.,
Girgaum, Mumbai 400004, Maharashtra
Tel : (022) 2385 6339 / 2386 9976, Fax : (022) 2386 9976
Email : niralimumbai@pragationline.com

➢ **DISTRIBUTION BRANCHES**

JALGAON

Nirali Prakashan : 34, V. V. Golani Market, Navi Peth, Jalgaon 425001,
Maharashtra, Tel : (0257) 222 0395, Mob : 94234 91860
Email : niralijalgoan@pragationline.com

KOLHAPUR

Nirali Prakashan : New Mahadvar Road, Kedar Plaza, 1st Floor Opp. IDBI Bank
Kolhapur 416 012, Maharashtra. Mob : 9850046155
Email : niralikolhapur@pragationline.com

NAGPUR

Nirali Prakashan : Above Maratha Mandir, Shop No. 3, First Floor,
Rani Jhanshi Square, Sitabuldi, Nagpur 440012, Maharashtra
Tel : (0712) 254 7129; Email : niralinagpur@pragationline.com

DELHI

Nirali Prakashan : 4593/15, Basement, Agarwal Lane, Ansari Road, Daryaganj
Near Times of India Building, New Delhi 110002 Mob : 08505972553
Email : niralidelhi@pragationline.com

BANGALURU

Nirali Prakashan : Maitri Ground Floor, Jaya Apartments, No. 99, 6th Cross, 6th Main,
Malleswaram, Bangaluru 560 003, Karnataka
Mob : +91 9449043034

Other Branches : Hyderabad, Chennai

niralipune@pragationline.com | www.pragationline.com

Also find us on www.facebook.com/niralibooks

Acknowledgement ...

I am grateful to Nirali Prakashan for providing me the opportunity to write a book on **Solid Waste Management** for diploma students. I must place on record my sincere appreciation and thanks to the scientific and technical institutions and various Ministries of the Government of India for using their web pages for the book.

I am also grateful to my PhD Thesis Supervisor Louis F Cohn, PhD, PE, FASCE; Professor of Civil and Environmental Engineering, University of Louisville, Kentucky, USA for providing me the inspiration to write books on Environmental Engineering stream to help society.

My acknowledgements wouldn't be complete without appreciating the immense encouragement that was provided to me during this project by respected Shri Dayanand Meshram, Joint Director, DTE, Mumbai and Dr. Hemant Taskar, Principal, Government Polytechnic, Mumbai. I thank them from the bottom of my heart for this encouragement.

I am deeply indebted to my wife Mrs. Sanjeeta Koul Gupta for her patience and support while I was busy with my work. Also thankful to my Son, Master Prinesh Bansal for his affection when I was writing the book.

Last but not the least, I am thankful to Mr. Dineshbhai Furia, Mr. Jignesh Furia, Mr. Shashikant Patel, Mr. Paras, Mrs. Manasi Pingle, Mr. Malik Shaikh, Mrs. Anjali Muley and the staff of Nirali Prakashan for helping me to come out successful with this project in limited time.

Place: Mumbai
15th December, 2019

Dr. Dinesh Kumar Gupta, BE (NIT), MTech (IIT),
PhD (USA), PE (Kentucky), MASCE, Program Evaluator (PEV), ABET,
Environmental Consultant, Lecturer, Government Polytechnic, Mumbai
Email: dinesh3912@hotmail.com

Preface ...

It gives us immense pleasure to present this book of **Solid Waste Management** for Third Year, Semester VI, Diploma students in Civil Engineering (Elective), strictly written as per the new MSBTE's 'I' scheme syllabus. This book was motivated by the desire we and others including MSBTE had to further the evolution of the specialized courses in Civil Engineering.

This book deals with solid waste which is the unwanted or useless solid materials generated from combined residential, industrial and commercial activities in a given area. It may be categorized according to its origin (domestic, industrial, commercial, construction or institutional); according to its contents (organic material, glass, metal, plastic paper etc); or according to hazard potential (toxic, non-toxin, flammable, radioactive, infectious etc). Management of solid waste reduces or eliminates adverse impacts on the environment and human health and supports economic development and improved quality of life. A number of processes are involved in effectively managing waste for a municipality. These include monitoring, collection, transport, processing, recycling and disposal.

The material is presented such that most chapters need not be read in any particular sequence. However, if a formal class is based on the subject, the chapters should be covered in sequential order. We have made every effort to make the subject easy, interesting and meaningful and have avoided unnecessary and excessive material.

Constructive criticism and suggestions are more than welcome and shall highly be appreciated. We solicit your views for improvement on the following e-mail address:

Dr. Dinesh Gupta - dinesh3912@hotmail.com

V. K. Sonarkar - vksonarkar@gmail.com

Authors

Syllabus ...

1. Fundamentals of Solid Waste Management **(Marks 06, Hours 04)**

1.1 Definition of Solid Waste

1.2 Meaning of Different Solid Waste – Domestic Waste, Commercial Waste, Industrial Waste, Market Waste, Agricultural Waste, Biomedical Waste, E-waste, Hazardous Waste, Institutional Waste

1.3 Sources of Solid Waste, Classification of Solid Waste – Hazardous and Non-hazardous Waste.

1.4 Physical and Chemical Characteristics of Municipal Solid Waste.

1.5 Impact of Solid Waste on Environment.

1.6 Solid Waste Management Techniques – Solid Waste Management Hierarchy, Waste Prevention and Waste Reduction Techniques.

1.7 Factors Affecting the Solid Waste Generation.

2. Storage, Collection and Transportation of Municipal Solid Waste

(Marks 14, Hours 10)

2.1 Storage of Solid Waste

2.2 Collection Methods of Solid Waste

2.3 Tools and Equipment - Litter Bin, Broom, Shovels, Handcarts, Mechanical Road Sweepers, Community Bin - like Movable and Stationary Bin

2.4 Transportation of Municipal Waste

2.5 Transportation Vehicles with their Capacity, Working - Animal Carts, Auto Vehicles, Tractors or Trailers, Trucks, Dumpers, Compactor Vehicles. Transfer Station - Meaning, Necessity, Location

2.6 Role of Rag Pickers and their Utility for Society

2.7 Organization Pattern of Solid Waste Management System, Practices According to Population of the Town or City

3. Disposal of Municipal Solid Waste **(Marks 16, Hours 14)**

3.1 Concept of Composting of Waste, Principles of Composting Process. Factors Affecting the Composting Process

3.2 Methods of Composting

(A) Manual Composting – Bangalore Method, Indore Method

(B) Mechanical Composting – Dano Process

(C) Vermicomposting

3.3 Land Filling Technique, Factors to be Considered for Site Selection

- Land Filling Methods - Area Method, Trench Method and Ramp Method.
- Leachate and Its Control, Biogas from Landfill
- Advantages and Disadvantages of Landfill Method
- Recycling of Municipal Solid Waste

3.4 Incineration of Waste

- Introduction of Incineration Process
- Types of Incinerators - Flash, Multiple Chamber Incinerator
- Products of Incineration Process with their Use
- Pyrolysis of Waste - Definitions, Methods

3.5 Product of Incineration Process

- Advantages and Disadvantages of Incineration Process

4. Biomedical Waste Management and Health Aspects and Public Involvement in Solid Waste Management (Marks 14, Hours 08)

Biomedical Waste Management

4.1 Definition of Bio Medical Waste.

4.2 Sources and Generation of Biomedical Waste

4.3 Classification of Biomedical Waste

4.4 Management Technologies

Health Aspects and Public Involvement in Solid Waste Management

4.4 Health Aspects During Handling and Processing.

4.5 Health Problems During Time of Segregation, Recovery, Recycling and Reuse of Solid Waste.

4.6 Public Involvement and Participation in Solid Waste Management Practices.

5. Industrial Waste Management and E-waste Management (Marks 14, Hours 08)

Industrial Waste Management :

5.1 Variety of Industrial Waste

5.2 Collection and Disposal of Industrial Waste

5.3 Control Measures for Industrial Waste

5.4 Recycling of Industrial Waste

E-waste Management :

5.5 Definition of E-waste, Varieties of E-wastes, Dangers of E-waste

5.6 Recycling of E-waste

5.7 Disposal of E-waste

6. Legal Aspects of Solid Waste Management (Marks 06, Hours 04)

Legal Aspects :

6.1 Legal Aspects - Present Scenario

6.2 Municipal Solid Waste Management Rules, 2016

6.3 Biomedical Waste Management Rules, 2016

6.4 E-Waste Management Rules, 2016

6.5 Construction and Demolition Waste Management Rules, 2016

6.6 Hazardous and Other Wastes Management Rules, 2016

6.7 Plastic Waste Management Rules, 2016

6.8 Role of Central Pollution Control Board and Maharashtra Pollution Control Board in Management of Solid Waste from Various Sources.

***

Contents ...

5. Industrial Waste Management and E-waste Management 5.1 - 5.18

6. Legal Aspects of Solid Waste Management 6.1 - 6.14

***

Chapter **1...**

Fundamentals of Solid Waste Management

Weightage of Marks = 06, Teaching Hours = 04

Syllabus

1.1 Definition of Solid Waste

1.2 Meaning of Different Solid Waste – Domestic Waste, Commercial Waste, Industrial Waste, Market Waste, Agricultural Waste, Biomedical Waste, E-waste, Hazardous Waste, Institutional Waste

1.3 Sources of Solid Waste, Classification of Solid Waste – Hazardous and Non-hazardous Waste.

1.4 Physical and Chemical Characteristics of Municipal Solid Waste.

1.5 Impact of Solid Waste on Environment.

1.6 Solid Waste Management Techniques – Solid Waste Management Hierarchy, Waste Prevention and Waste Reduction Techniques.

1.7 Factors Affecting the Solid Waste Generation.

About this Chapter

At the end of this chapter, students will be able to:

- Explain the principles of waste reduction in the given condition.
- Classify the given solid wastes according to their sources.
- Describe the characteristics of the given solid wastes.
- Justify the need of solid waste management in the given situation.
- Assess the impact of solid waste management on the environment in the given situation.
- List the factors generating solid wastes in the given specific area with justification.

1.0 OVERVIEW

- For many years in the recent past, people have been shifting from rural and semi-urban areas to towns and cities. The proportion of population residing in urban India has increased from 10.84% in 1901 to 25.70% in 1991.
- The number of class I cities has increased from 212 to 300 during 1981 to 1991, while class II cities have increased from 270 to 345 during the same period. The increase in the population in class I cities is quite high as compared to that in class II cities.
- The uncontrolled growth in cities has created the problems in areas such as water supply, sewerage and municipal solid waste management.
- It is estimated that the total solid waste generated by 300 million people living in urban India is 38 million tonnes per year.

- About 1,00,000 MT of Municipal Solid Waste is generated daily in the country.
- Per capita waste generation in major cities ranges from 0.2 kg to 0.6 kg.
- Generally, the collection efficiency ranges between 70 to 90% in major metro cities whereas in several smaller cities the collection efficiency is below 50%.
- It is also estimated that, the Urban Local Bodies spend about ₹ 500 to ₹ 1500 per tonne on solid waste for collection, transportation, treatment and disposal.
- Mumbai generates waste to the tune of approximately 7,025 tonnes per day. The waste consists of:
  - 5,025 tonnes of mixed waste (bio-degradable and recyclable).
  - 2,000 tonnes of debris and silt.
- The biodegradable waste (wet waste) is made up of vegetable and fruit remainders, leaves, spoiled food, eggshells, cotton etc. Recyclable waste (dry waste) consists of newspapers, thermocol, plastic, battery cells, wires, iron sheets, glass etc. Debris includes construction waste, renovation waste, demolition waste etc. Silt comprises earth and clay from drains and road corners.

1.1 DEFINITION OF SOLID WASTE (W-16)

- Solid waste is generally defined as 'non-soluble material that is thrown away in a solid or semi-solid form'. This includes garbage, refuse, sludge and other waste domestic materials, as well as waste from industrial, commercial, agricultural and mining operations.

1.2 MEANING OF DIFFERENT SOLID WASTE (S-19)

(a) Domestic waste: (W-16)

- Domestic waste also known as "municipal solid waste", is the waste that is generated as a result of the ordinary day-to-day activities at homes and is:
  (i) either taken from the homes by or on behalf of the person who generated the waste; or
  (ii) collected by or on behalf of a local government/bodies as part of a waste collection and disposal system.

(b) Commercial waste: (W-16)

- Commercial waste can be defined as 'any waste generated as a result of carrying out a business or commercial activity, including associated lawn and garden trimmings (cut grass, leaves etc.) from normal maintenance of the business premises'. Commercial waste also includes rubbish produced by the customers i.e. food wrappers and containers.

(c) Industrial waste: (S-16)

- Industrial waste is the waste produced by industrial activity which includes any material that becomes useless during a manufacturing process such as that of factories, mills and mines. Also industrial solid waste means solid waste generated by manufacturing or industrial processes and it normally does not include the hazardous waste as it is a separate category of solid waste.

(d) Market waste:

- Market waste is that waste which is generated out of marketing activities. It is primarily organic waste, such as leaves, skins and unsold food, discarded at or near food markets.

(e) Agricultural waste: (W-16)

- Agricultural waste, which includes both natural (organic) and non-natural wastes, is a general term used to describe waste produced on a farm by various farming activities. These activities may include dairy farming, horticulture, seed growing, livestock breeding, grazing land, market gardens, nursery plots and woodlands.

(f) Biomedical waste: (S-16)

- Biomedical waste is that waste which is of medical or medical laboratory origin (e.g., packaging, unused bandages, infusion kits etc.), as well research laboratory waste containing biomolecules or organisms that are restricted to be released to environment.
- Discarded needles are also considered as biomedical waste whether they are contaminated or not, due to the possibility of being contaminated with blood and their ability to cause injury when not properly enclosed and disposed off.

(g) E-waste:

- E-waste is that waste which consists of discarded electrical or electronic devices. All electronic scrap components, such as CRTs, may contain contaminants such as lead, cadmium, beryllium, or brominated flame retardants.
- Even in developed countries recycling and disposal of e-waste may involve significant risk to workers and communities and great care must be taken to avoid unsafe exposure in recycling operations and leaking of materials such as heavy metals from landfills and incinerator ashes.

(h) Hazardous waste: (W-16)

- Hazardous waste is the waste that can have substantial or potential threats to public health or the environment. These wastes are either flammable (or ignitable), reactive, corrosive or toxic or a combination of these. For example, dry cleaners, automobile repair shops, hospitals, slaughterers, and photo processing centers may all generate hazardous waste.
- Some hazardous waste generators are larger companies such as chemical manufacturers, electroplating companies and oil refineries.

(i) Institutional waste:

- Waste generated at institutions such as schools, libraries, hospitals, prisons etc. is known as institutional waste. It may include metal and glass pieces, plastic, yard waste, food waste, corrugated containers, paper and card board waste.

(j) Construction and demolition wastes:

- Construction and demolition wastes are the waste materials generated by the construction, refurbishment, repair and demolition of houses, commercial buildings and other structures. It mainly consists of earth, stones, concrete, bricks, lumber, roofing materials, plumbing materials, heating systems and electrical wires, and parts of the general municipal waste stream. But when generated in large amounts at building and demolition sites, it is generally removed by contractors for filling low lying areas and by urban local bodies for disposal at landfills.

(k) Sewage wastes:

- The solid by-products of sewage treatment are classified as sewage wastes. They are mostly organic and derive from the treatment of organic sludge from both the raw and treated sewage.
- The inorganic fraction of raw sewage such as grit is separated at the preliminary stage of treatment, but because it contains putrescible organic matter which may have pathogens, must be buried/disposed off without delay. The bulk of treated, dewatered sludge is useful as a soil conditioner but its use for this purpose is uneconomical. The solid sludge therefore enters the stream of municipal wastes unless special arrangements are made for its disposal.

1.3 SOURCES OF SOLID WASTE (S-15, 16, 19; W-16, 18)

Various Sources of Solid Waste:

Following are the sources of solid waste.

(i) Residential: This includes wastes from residential houses like dwellings, apartments etc. and consist of leftover food, vegetables peels, plastic, clothes, ashes etc.

(ii) **Commercial :** This refers to waste generated from restaurants, hotels, motels, stores, markets, auto-repair shops, medical facilities etc. and consist of leftover food, glasses, metals, ashes etc.

(iii) **Institutional :** This includes waste coming from educational, administrative and public buildings like prisons etc. and consist of paper, plastic, glasses etc.

(iv) **Municipal :** This mainly refers to waste from various municipal activities like construction and demolition, street cleaning, landscaping etc. and consist of leaf matter, dust, building debris, waste water treatment plant, residual sludge etc.

(v) **Industrial :** This mainly consists of waste generated from various industrial activities and consist of process wastes, ashes, construction and demolition wastes, hazardous wastes etc.

(vi) **Agricultural :** This includes wastes coming from fields, orchards, vineyards, farms etc. and consist of agricultural remains, spoiled vegetables and grains, litter etc.

(vii) **Open areas :** This mainly refers to waste from streets, alleys, parks, playgrounds, beaches, highways, recreational areas etc.

Types of Solid Waste: **(S-15, 19)**

1. Municipal waste
2. Hazardous waste
3. Bio-medical waste
4. Electronic waste
5. Industrial Waste

1.4 CLASSIFICATION OF SOLID WASTE

- Solid waste can be classified based on its source as residential, agricultural, commercial, industrial, institutional or healthcare waste. It can also be classified as hazardous or non-hazardous and also as biodegradable or non-biodegradable. Another word used in the context of biodegradable solid waste is putrescible, which means the waste which decomposes (rots down) quite quickly. Waste can also be classified as combustible or non-combustible depending on whether it will burn or not.

Table 1.1: Classification of Solid Waste **(W-15)**

Main classification	Type	Short description	Examples
Hazardous waste	Solid or semi-solid	Substances that are either ignitable, corrosive, reactive, infectious or toxic.	Some obsolete pesticides such as DDT, dieldrin etc.
Non-hazardous waste	Putrescible	Easily decomposable/ biodegradable solid waste.	Food wastes
	Non-putrescible	Non-biodegradable solid waste. It includes combustible and non-combustible waste.	Plastic bags, bottled water containers, tin cans etc.

1.4.1 Hazardous Waste

- It can be solid or semisolid and is either ignitable, corrosive, reactive, infectious or toxic or a combination of these. Examples include pesticides such as DDT, dieldrin etc.

Characteristics of Hazardous Waste: **(W-16; S-15)**

1. **Ignitability:** Ignitable wastes create fires under certain conditions or are spontaneously combustible, or have a flash point less than 60°C (140°F).
2. **Corrosivity:** Corrosive wastes are acids or bases (pH less than or equal to 2 or greater than or equal to 12.5) that are capable of corroding metal containers, such as storage tanks, drums and barrels.
3. **Reactivity:** Reactive wastes are unstable under "normal" conditions. They can cause explosions, toxic fumes, gases, or vapours when mixed with water.

4. **Toxicity:** Toxic wastes are harmful or fatal when ingested or absorbed (e.g., containing mercury, lead etc.). When toxic wastes are disposed off on land, contaminated liquid may drain (leach) from the waste and pollute ground water. Toxicity is defined through a laboratory procedure called the *Toxicity Characteristic Leaching Procedure.*

1.4.2 Non-hazardous Waste

- Non-hazardous waste is that waste which is not substantial or potential threat to public health or the environment. It can further be classified as –
  - **(a) Putrescible:** It is that waste which is easily decomposable/biodegradable solid waste. Example: Food waste.
  - **(b) Non-putrescible:** It is that waste which is non-biodegradable solid waste, it includes combustible and non-combustible waste. Example: Plastic bags, bottled water containers, tin cans etc.

1.4.3 Difference between Hazardous and Non-hazardous Solid Waste (S-18)

Sr. No.	Hazardous waste	Non-hazardous
1.	Waste which is either ignitable, corrosive, reactive, infectious or toxic or combination of these is called as hazardous waste.	Waste which is not potential threat to public health and environment is called as non-hazardous waste.
2.	Requires special management considerations, because the treatment method for one of the hazards may be inappropriate for the treatment of another.	Simple treatment and disposal methods. Can be used for all types of non-hazardous waste like landfilling.
3.	Cost of treatment and disposal of waste is more.	Treatment and disposal cost is less.
4.	Examples: Industrial waste, E-Waste, Bio-medical waste.	Examples: Food waste, plastic bags, tin cans etc.

Table 1.2: Hazardous and Non-hazardous Substances, their Origin and Health Effects (S-19)

Substances	Origin	Health effects
Hazardous Substances:		
Lead	Solder in PCB (printed circuit board).	Damage to nervous system, kidney, brain development of children.
Mercury	Tubes, PCB, Thermostats.	Chronic damage to brain, skin disorder, memory loss, muscle weakness.
Cadmium	Chip resistor and semiconductor.	Kidney and lever damage, neural damage
Hexavalent chromium	Corrosion protection of steel plates.	Asthamatic bronchitis, DNA damage.
Barium	Front panel of CRT.	Muscle weakness, damage to heart/ lever/spleen.
Beryllium	Motherboard.	Carcinogenic, skin disease, beryllicosis.
Plastic	Cabling/Computer housing.	Immune system damage, hormone interference.

Substances	Origin	Health effects
Non-hazardous Substances:		
Tin	Obtained from the mineral cassiterite.	Tin dust can irritate the skin and delicate tissue.
Copper	Ore of cuprite.	Long-term exposure to copper can cause irritation of the nose, mouth and eyes and it causes headaches, stomachaches, dizziness, vomiting and diarrhoea.

Aluminium, Iron, Germanium, Silicon, Nickel, Lithium, Zinc, Gold are some more non-hazardous substances.

Table 1.3: Classification of Solid Waste with Examples **(W-15, 18)**

Type	Examples
Food waste (Garbage)	Waste from preparation, cooking and serving of food market refuse, waste from handling, storage and sale of meat and vegetables.
Rubbish	Combustible (primarily organic) paper, cardboards, cartons, wood boxes, plastics, rags, clothes, beddings, leather, rubber, grass, leaves, yard trimmings. Non-combustible (primarily inorganic) metals, tin cans, metal foils, dirt, stones, bricks, ceramics, crockery, glass bottles, other mineral refuse.
Ashes and residues	Residues from fires used for cooking and for heating buildings, cinders, clinkers, thermal power plants.
Bulky waste	Large auto parts, tyres, stoves, refrigerators, other large appliances, furniture, large crates, branches of trees etc.
Street waste	Street sweepings, dirt, leaves, catch basin dirt, animal droppings, content of litter, receptacles, dead animals.
	Dead animals: Small animals: cats, dogs, poultry etc. Large animals: Horses, cows etc.
Construction and demolition waste	Plumber, roofing and sheathing scrap, rubble, broken concrete, plaster, conduit pipes, insulating wires etc.
Industrial waste and sludges	Solid wastes resulting from industry processes and manufacturing operations, such as food processing wastes, boiler house cinders, wood, plastic and metal scraps and shavings etc., sludge of sewage treatment plants and septic tanks, coarse screenings, grit etc.
Hazardous waste	Pathological waste, explosives, radioactive material, toxic waste etc.
Horticulture waste	Tree trimmings, leaves, waste from parks and gardens etc.

1.5 PHYSICAL AND CHEMICAL CHARACTERISTICS OF MUNICIPAL SOLID WASTE

- National Environmental Engineering Research Institute (NEERI) has carried out extensive studies on Physical and Chemical characteristics of solid waste from 43 cities during 1970-1994.

1.5.1 Physical Characteristics (S-15; W-16)

- Density, moisture content, size of waste constituents, specific weight or weight of waste per unit volume, particle size and distribution, field capacity/percentage of moisture in wet solid and calorific value are the main physical characteristics of solid waste.
  1. **Density:** 50-290 kg/m^3. Knowledge of the density of waste i.e. mass per unit volume is essential for the design of all elements of the solid waste management system.
  2. **Moisture content:** 05-60%. Moisture content is a critical determinant in the economic feasibility of waste treatment by incineration, since energy must be supplied for evaporation of water and in raising the temperature of the water vapour.
  3. **Size distribution:** 1 cm to 50 cm. The measurement of the size distribution of particles in the waste stream is important because of its significance in the design of mechanical separators and shredders.
  4. **Field capacity:** Field capacity is defined as 'total amount of moisture held in a waste matter under gravity'. It is important as it controls the rate of leachate generation and permeability.
  5. **Permeability of compacted waste:** The hydraulic conductivity of compacted waste is an important physical property to a large extent which governs movement of liquid and gases in landfill.
  6. **Porosity:** It represents the amount of voids per unit overall volume of material.
- The calorific value ranges between 800-1000 kcal/kg.
- The paper content generally varies between 2.9 to 6.5% and increases with the increase in population.
- The plastics, rubber and leather contents are lower than the paper content, and do not exceed 1% except in metropolitan cities.
- The metal content is also low, viz. less than 1%. The low values are essentially due to the large scale recycling of these constituents. The paper is recycled on a priority basis while the plastics and glass are recycled to a lesser extent.
- The biodegradable fraction is quite high, essentially due to the habit of using fresh vegetables in India. The high biodegradable fraction also makes it necessary for frequent collection and removal of solid waste from the collection points.
- The ash and fine earth content of Indian municipal solid waste is high due to the practice of inclusion of the street sweepings, drain silt, and construction and demolition debris in municipal solid waste. The proportion of ash and fine earth reduces with increase in population due to improvements in the road surfaces.
- High ash and earth content increases the densities of municipal solid waste which are between 350 and 550 kg/m^3 in Indian cities.

1.5.2 Chemical Characteristics (W-15; S-19)

- A knowledge of chemical characteristics of waste is essential in determining the ability of any treatment process. Chemical characteristics include (i) Chemical; (ii) Bio-chemical; and (iii) Toxic.
- Chemical characteristics include pH, nitrogen, phosphorus and potassium (N-P-K), total carbon, C/N ratio, calorific value.
- The chemical characteristics indicate that the organic content of the samples on a dry weight basis ranges between 20 to 40%.
- The nitrogen, phosphorus and potassium content of the municipal solid waste ranges between 0.5 to 0.7%, 0.5 to 0.8% and 0.5 to 0.8% respectively.
- Knowledge of the chemical characteristics is essential in selecting and designing the waste processing and disposal facilities.

- Ragpickers are observed to be more active in bigger cities. They prefer to remove paper, plastics, rags and packaging and such other material, which is light and also have a high calorific value.
- The remaining waste hence tends to have a higher inert content and a lower calorific value. The demolition activity is observed to increase with population leading to increased inert content and reduced organic content in MSW.
- Bio-chemical characteristics include carbohydrates, proteins, natural fibre and biodegradable factor.
- Toxicity characteristics include heavy metals, pesticides, insecticides, Toxicity test for leachates (TCLP) etc.

Some Characteristics of Solid Waste : **(W-18)**

1. **pH:** The pH of fresh solid waste is normally around 7. During decomposition, it tends to become acidic, and stabilized solid waste has an alkaline pH.
2. **Organic content:** The knowledge of organic content helps to assess the feasibility of biological processing - composting and anaerobic digestion.
3. **Carbon content:** The carbon is determined by using the New Zealand formula in which the percent organic matter is divided by 1.724.
4. **Nitrogen content:** Nitrogen, phosphorous and potash values are important in composting.
5. **Toxicity:** Toxicity characteristics include heavy metals, pesticides, insecticides etc.

1.5.3 Proximate Analysis of Solid Waste (S-15)

- Determination of fixed carbon, volatile combustible matter, moisture and ash content of the waste in order to estimate its capability as a fuel is called proximate analysis of solid waste.
- The fixed carbon, volatile combustible matter can be burnt while moisture and ash not. The vapourization of the moisture consumes heat.

Method of Analysis (Tests):

1. **Moisture:** Determination from the loss of weight by heating at 105°C for one hour.
2. **Volatile combustible matter:** The additional loss of weight after ignition at 950°C in a covered crucible (O_2 is excluded).
3. **Fixed carbon:** Combustible residue after the volatile combustible matter is removed; ignition at 600 to 900°C.
4. **Ash:** The weight of residue after combustion in an open crucible.

   % fixed carbon = 100 % – % moisture – % ash – % volatile matter. It does not provide any information of possible pollutants emitted during combustion. These data are determined by ultimate analysis.

1.6 IMPACT OF SOLID WASTE ON ENVIRONMENT (S-16; W-15, 16, 18)

1. Ground water contamination by the leachate generated by the illegal waste dump. Surface water contamination by the run-off from the waste dump.

   Leaching is a process by which solid waste enter soil and ground water, and contaminate them and pollute water bodies, or ground water systems.
2. Bad odour, pests, rodents and wind-blown litter in and around the waste dump. Generation of inflammable gas (e.g. methane) within the waste dump
3. Bird menace above the waste dump which affects flight of aircraft.
4. Fires within the waste dump.
5. Erosion and stability problems relating to slopes of the waste dump.
6. Epidemics through stray animals.
7. Acidity to surrounding soil and release of green house gas.
8. Waste breaks down in landfills to form methane, which causes greenhouse gas.

   Carbon dioxide and Methane produced from solid waste are extremely harmful to the environment.

9. Change in climate and destruction of ozone layer due to waste biodegradable.
10. Open air dumping creates unhygienic atmosphere and poses enormous threat to the people. Causes aesthetic problem and nuisance due to nauseating pungent odour.
11. Promotes spreading of diseases.
13. The situation is more aggravated by the indiscriminate disposal of hospital and clinical waste.
12. Presence of extremely high level of total and facial *E-coli* form.

1.6.1 Solid Waste as a Global Problem for Modern World (S-18)

"Solid waste is creating a global problem for modern world".

- Around the world, waste generation rates are rising. In 2012, the worlds' cities generated tons of solid waste per year, amounting to a footprint of 1.2 kilograms per person per day.
- With rapid population growth and urbanization, municipal waste generation is expected to rise to 2.2 billion tones by 2025.
- Compared to those in developed nations, residents in developing countries, especially the urban poor, are more severely impacted by unsustainably managed waste.
- In low and middle-income countries, waste is often disposed in unregulated dumps or burned openly. These practices create serious health, safety and environmental consequences.
- Poorly managed waste serves as a breeding ground for disease vectors, contributes to global climate change through methane generation, and even promotes urban violence.
- Managing waste properly is essential for building sustainable and livable cities, but it remains a challenge for many developing countries and cities. Effective waste management is expensive, often comprising 20%-50% of municipal budgets.
- Operating this essential municipal service requires integrated systems that are efficient, sustainable, and socially supported.

1.7 SOLID WASTE MANAGEMENT TECHNIQUES (S-16; W-18)

- Waste management techniques in accordance with the waste hierarchy include waste prevention/reduce at source, reuse, recycle, recovery and disposal of solid waste with prevention at source to be given first preference to the disposal being the last preferred alternative.

1.7.1 Solid Waste Management Hierarchy (4R of Solid Waste Management Hierarchy) (S-15; W-15, 18)

The main aim of waste hierarchy is to generate minimum amount of waste and obtain maximum benefits from products.

- Following are the various stages in SWM hierarchy:
    1. **Prevention:** Preventing the use of such raw material in production which produces maximum solid waste and selecting the alternative raw materials.
    2. **Minimization:** If such alternative raw materials are less possible then minimize the use of raw materials producing more waste by implementing different techniques.

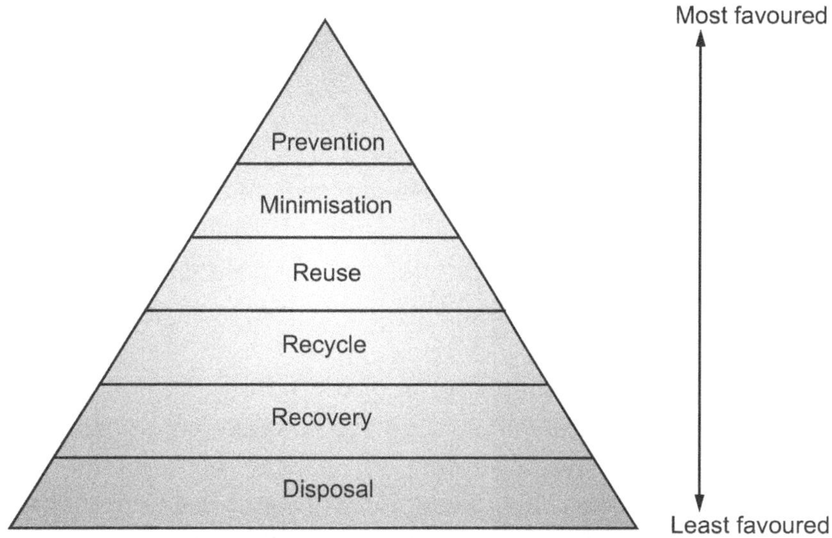

Fig. 1.1: Solid Waste Management Hierarchy

3. **Reuse:** It is the next desirable option in which some materials are repeatedly used again and again for same purpose.
4. **Recycle:** In this stage, collection, sorting of recyclable products is done and then they are manufactured into new products.
5. **Recovery:** In this stage, the recoverable materials are processed which includes activities like recycling and composting.
6. **Disposal:** It is the last option and should be considered after all other possible actions to recover that waste matter. It may includes incineration, dumping.

- The waste hierarchy refers to the four 'R's of reduce, reuse, recycle, recovery and disposal. The waste hierarchy has taken many forms over the past decade, but the basic concept has remained the cornerstone of most waste minimization strategies.
- The aim of the waste hierarchy is to extract the maximum practical benefits from products and to generate the minimum amount of waste. Suitability or goodness of option varies from top to bottom in decreasing order; top option being the best and the bottom most being the least preferred.

Four R's in Waste Hierarchy:

Following are the four R's in waste hierarchy.

(i) **Reduce:** The concept of reducing is what is produced and what is consumed i.e. multiuse items is essential to reduce the waste e.g. print on both side of paper to reduce paper wastage. Reducing waste is minimizing waste quantity.

(ii) **Reuse:** Reuse or re-purpose items for different functions than what they are intended. e.g. Old newspaper can be used to make packing items.

(iii) **Recycle:** Recycle means it will be transformed again into a raw material that can be shaped into a new item. e.g. glass, metals etc.

(iv) **Recover:** It is the fourth option in which recoverable materials are processed and includes activities like recycling and composting.

1.7.1.1 Reduction of Waste and Waste Prevention (S-16; W-16)

Waste prevention:

- Waste prevention means reducing waste by not producing it. It is also known as source reduction i.e. reducing the waste stream is the most significant of all the options to manage waste. If we never generate the waste, then we do not have to invent ways to dispose it off.
- Waste prevention is the preferred waste management activity as it actually avoids waste generation.
- Waste prevention conserves resources, protects the environment, and prevents the formation of greenhouse gases.

Waste reduction:

- To reduce the waste we usually have to make significant lifestyle changes. Reduce the amount we buy in the first place. Purchase only the amount we need. By becoming better environmental shoppers, we can reduce the waste we generate.
- Waste reduction can be achieved in three ways:
  (i) By eliminating the need of the product.
  (ii) By reducing the amount of material used per product without sacrificing the utility of that product.
  (iii) By increasing lifetime of a product.
- To reduce waste we usually have to make significant lifestyle changes.
- Office paper wastage can be reduced by implementing a formal policy to duplex all draft reports and by making training of manuals and personnel information available electronically.
- Use less material and improve product design.
- Switch to reusable things.

1.7.1.2 Reuse of Waste

- The idea of being wasteful makes many people uncomfortable. Yet most of us continue to waste because we cannot think of anything better to do with the used, old, empty, worn, broken, ugly things that are of no use.
- We are conditioned to think of things that are useless; so we throw them away without much thought. Most people buy far more than they can use effectively as evidenced by bulging terraces, attics and garages.
- The process of reusing is started with the assumption that, the used materials that flow through our lives can be a resource rather than refuse. Waste, after all, is in the eye of the beholder. What is one person's trash is another one's treasure. So our trace can be very useful to others or less privileged.
- If we really look at things we are throwing away, we can learn to see them as materials that can be reused to solve everyday problems and satisfy everyday needs. Most of us, however, have not even begun to exploit the resources in our trash. Once we have our mind set that we can use trash for positive uses, we can begin to brainstorm and generate ideas. Reusing saves money, conserves resources and it satisfies the human urge to make things.

1.7.1.3 Recycle of Waste (W-18; S-19)

Definition:

The action or process of converting waste into reusable material is called recycling.

- Recycling turns materials that would otherwise become waste into valuable resources and generates a combination of environmental, financial and social benefits. After collection, materials (e.g., glass, metal, plastics and paper) are separated and sent to facilities that can process them into new products and materials.

Purpose of Recycling:

(i) To save raw material resources in production.
(ii) To save energy in production.
(iii) To reduce solid waste in landfills.
(iv) To minimize environmental pollution.
(v) To reduce the consumption of conventional raw materials.
(vi) To reduce air pollution from incinerations.

Status of Recycling of Solid Waste in India :

- The rules in India require municipalities to ensure community participation in waste segregation and to promote recycling of segregated materials.
- In India there is no formal recycling system but informal recyclers are there and play an important role in SWM.
- These comprise of unorganized and unrecognized establishments which are difficult to be monitored by government agencies.
- However resource recovery through material recycling is taking place in India in a big way, through unorganized ways.

1.7.1.4 Recovery from Waste

- Energy recovery from waste is the conversion of non-recyclable waste materials into useable heat, electricity, or fuel through a variety of processes, including combustion, gasification, pyrolization, anaerobic digestion and landfill gas (LFG) recovery. This process is often called waste-to-energy (WTE).
- At an MSW combustion facility, MSW is unloaded from collection trucks and placed in a trash storage bunker. An overhead crane is used to sort the waste and then lift it into a combustion chamber to be burned. The heat released from burning is used to convert water to steam. The

steam is then sent to a turbine generator to produce electricity. The remaining ash is collected and taken to a landfill. Particulates are captured by a high-efficiency bag house (a filtering system). As the gas stream travels through these filters, more than 99% of particulate matter is removed. Captured fly ash particles fall into hoppers (funnel-shaped receptacles) and are transported by an enclosed conveyor system to the ash discharger where they are wetted to prevent dust and mixed with the bottom ash from the grate. The ash residue is then conveyed to an enclosed building where it is loaded into covered, leak-proof trucks and taken to a landfill designed to protect against groundwater contamination. Ash residue from the furnace can be processed for removal of recyclable scrap metals.

1.7.1.5 Disposal of Waste

- Last technique, if other alternatives as per the above mentioned hierarchy are exhausted, to get rid of solid waste, is the successful disposal of the solid waste by any one of the following methods:

  (i) Land Filling (ii) Composting (iii) Incineration

1.7.2 Types of Solid Waste from Engineering and Chemical Type Industries

(S-17)

Pulp and Paper Industry:

Sr. No.	Types of Solid Waste	Characteristics of Solid Waste
1.	Grit	Oil and grease
2.	Fly ash	Fine dust
3.	Sludge	High solids
4.	Bleached pulp	High pH

Chemical Industry:

Sr. No.	Types of Solid Waste	Characteristics of Solid Waste
1.	Slag	Acidic or alkaline
2.	Mud	High suspended solid
3.	Residues of refining	High alkalinity
4.	Scraps	Metal parts

1.8 FACTORS AFFECTING SOLID WASTE GENERATION (S-16, 19; W-15, 16)

Following are the factors which affect the solid waste generation:

(i) Population
(ii) Urbanization
(iii) Industrialization
(iv) Life style
(v) Family income
(vi) Size of family
(vii) Climatic condition of the area
(viii) Moisture content
(ix) Tourist number
(x) Habits and culture of people
(xi) Collection frequency
(xii) Public attitudes
(xiii) Source reduction/recycling.

1.9 WASTE MINIMIZATION APPROACH

- **Optimization of resources:** Waste reduction at individual and institutional level goes side by side with the utilization of raw materials.
- **Using again the scrap material:** This is the process in which individual and industry reuse the waste material as soon as it is produced. This keeps it from becoming a waste material.

- **Quality control improvement and process monitoring:** This technique is to ensure that, products produced are kept from rejection and this is increased by the inspection of frequency and monitoring point's inspection.
- **Exchanging waste:** This is the technique in which the waste product, which comes out of a process, becomes a raw material for another process. This is another way for reducing waste.
- **Point of use from ship:** To maintain and making deliveries for the raw materials to be used with the manufacturing process, at the point of assembly with fewer packages and wrappings can save from the waste production.
- **Zero waste:** This is a whole systems approach that aims to eliminate waste at the source and at all points down the supply chain, with the intention of producing no waste. It is a design philosophy which emphasizes waste prevention as opposed to end of pipe waste management

Images of Solid Waste:

Fig. 1.1

Fig. 1.2

Fig. 1.3

1.10 IMPORTANCE OF SOLID WASTE MANAGEMENT IN MAKING ANY CITY AS SMART CITY (W-18)

1. **Environmental friendly:**
   (a) Correct ways of waste management especially industrial waste, can keep our environment free from pollution.

   Example:

   (i) Heavy metal such as mercury which can harm to our ecosystem. Proper disposal is required.

   (ii) Nuclear waste management to prevent radiation to our natural environment.

2. **Waste causes resource depletion:**
   This is due to the common buying pattern: buy, throw, and then buy again. As the waste piles up high, the demand for more products also rises, almost exhausting the natural resources. This has a spiral effect, mainly involving threats to biodiversity, deforestation, pollution, and other environmental problems.

3. **Waste management:** (S-16)
   It is the collection, transport, processing, recycling or disposal, and monitoring of waste materials. The term usually relates to materials produced by human activity, and is generally undertaken to reduce their effect on health, the environment or aesthetics. Waste management is also carried out to recover resources from it. Waste management can involve solid, liquid, gaseous or radioactive substances, with different methods and fields of expertise for each.

Important Points

- In India, the increase in population in Class-I cities is high as compared to Class-II cities.
- The uncontrolled growth of population has created water supply, sewage and solid waste management problems.

- About 1,00,000 MT of Municipal Solid Waste is generated daily in India.
- Per capita waste generation is about 0.2 kg to 0.6 kg.
- Domestic waste, commercial waste, industrial waste, market waste, agriculture waste, biomedical waste, E-waste, construction and demolition waste are the types of waste.
- Residential, commercial, municipal services, industrial are the some important sources of solid waste.
- Solid waste can be classified as (a) Hazardous, (b) Non-hazardous.
- NEERI has carried out study of physical and chemical characteristics of solid waste.
- There is adverse effect of municipal solid waste on environment. So its proper management is required.
- Reduction of waste, recycle of waste, recovery from waste, disposal of waste are the some SWM techniques.
- Population, urbanization, industrialization, life style, size and income of family, habit and culture of people are some important factors which are responsible for generation of solid waste.

Practice Questions

1. Explain the following terms:
   (a) Domestic waste (b) Commercial waste (c) Industrial waste
   (d) Market waste (e) Agriculture waste (f) Biomedical waste
   (g) E-waste (f) Hazardous waste (g) Institutional waste
   (h) Construction and Demolition wastes (i) Sewage wastes
2. State the factors affecting the generation of waste.
3. Name the methods of disposal of solid waste.
4. Classify solid waste.
5. What is putrescible waste?
6. Describe physical characteristics of solid waste?
7. Describe chemical characteristics of solid waste?
8. Which is the best technique for handling solid waste?
9. Define solid waste.
10. What are four 'R's in waste hierarchy?

MSBTE Questions and Answers (As Per 'G' Scheme)

Summer 2016

1. Define the following: (i) E-waste, (ii) Biomedical waste. **(4M)**

Ans. Refer Section 1.2 (g) and 1.2 (f) respectively.

2. What are the sources of solid waste. **(4M)**

Ans. Refer Section 1.3.

3. State the factors affecting the solid waste generation. **(4M)**

Ans. Refer Section 1.8.

4. Draw the organization pattern of solid waste management. **(4M)**

Ans. Refer Section 1.7.1.

5. State the impact of solid waste on environment. **(4M)**

Ans. Refer Section 1.6.

6. Define Industrial waste and Biomedical waste. **(4M)**

Ans. Refer Section 1.2 (c) and (f).

7. Explain solid waste management techniques. **(4M)**

Ans. Refer Section 1.7.

8. Explain waste prevention and waste management. **(4M)**

Ans. Refer Section 1.7 and 1.10.

9. State hazardous solid waste. **(4M)**

Ans. Refer Sections 1.2 (h) and 1.4.1.

10. State the benefits of recycling of solid waste. **(4M)**

Ans. Refer Section 1.7.1.3.

Winter 2016

1. Define Solid waste and list any four sources of solid waste. **(4M)**

Ans. Refer Section 1.1 and 1.3.

2. State the various factors affecting on solid waste generation. **(4M)**

Ans. Refer Section 1.8.

3. Give the physical characteristics of solid waste. **(4M)**

Ans. Refer Section 1.5.1.

4. Give the impact of solid waste on environment. **(4M)**

Ans. Refer Section 1.6.

5. Give the meaning of following term: **(4M)**

   (i) Hazardous waste, (ii) Domestic waste, (iii) Agriculture waste, (iv) Commercial waste.

Ans. Refer Section 1.2 (h), (a), (e) and (b) respectively.

6. Explain the term waste prevention and waste reduction. **(4M)**

Ans. Refer Section 1.7.1.1.

7. State any four benefits of recycling of solid waste. **(4M)**

Ans. Refer Section 1.7.1.3.

Summer 2017

1. Define 'Solid Waste', also list different types of solid waste. **(4M)**

Ans. Refer Sections 1.1 and 1.2.

2. Explain the impact of Solid Waste on environment. **(4M)**

Ans. Refer Section 1.6.

3. Explain the various sources of Solid Waste. **(4M)**

Ans. Refer Section 1.3.

4. State four factors affecting the Solid Waste Generation. **(4M)**

Ans. Refer Section 1.8.

5. Define : **(4M)**

   (i) Biomedical Waste.

   (ii) E-Waste.

Ans. (i) Biomedical Waste: Refer Section 1.2 (f).

(ii) E-Waste: Refer Section 1.2 (g).

6. Explain the Hierarchy of Solid Waste Management. **(8M)**

Ans. Refer Section 1.7.1.

Winter 2017

1. Define 'solid waste'. **(4M)**

Ans. Refer Section 1.1.

2. List types of solid waste. **(4M)**

Ans. Refer Section 1.4.

3. List the characteristics of solid waste. Explain any two characteristics of solid waste along with solid waste management hierarchy. **(8M)**

Ans. Refer Sections 1.5 and 1.7.1.

4. State the sources of solid waste. Write impact of solid waste on environment. State the factors affecting generation of solid waste. **(8M)**

Ans. Refer Sections 1.3, 1.6 and 1.8.

5. State various ways of solid waste prevention and reduction. **(5M)**

Ans. Refer Section 1.7.1.1.

6. Define Biomedical waste. **(2M)**

Ans. Refer Section 1.2 (f).

7. Explain the following : **(4M)**

   (i) Reuse of solid waste, (ii) Purpose of recycling.

Ans. Refer Sections 1.7.1.2 and 1.7.1.3.

Summer 2018

1. State any eight impacts of solid waste on environment. **(4M)**

Ans. Refer Section 1.6.

2. State the various factors affecting solid waste generation. **(4M)**

Ans. Refer Section 1.8.

3. Define following terms: **(4M)**

   (i) Commercial waste, (ii) Industrial waste, (iii) Agricultural waste, (iv) Institutional waste.

Ans. Refer Section 1.2.

4. Differentiate between hazardous and non-hazardous solid waste. **(4M)**

Ans. Refer Sections 1.4.3 and Table 1.1.

5. "Solid waste is creating a global problem for modern world". Justify the statement. **(4M)**

Ans. Refer Section 1.6.1.

6. Explain the four "R's" of solid waste management hierarchy. **(4M)**

Ans. Refer Section 1.7.1.

Winter 2018

1. State types of solid waste and give examples of each. **(4M)**

Ans. Refer Section 1.2.

2. Explain impact of solid waste on environment. **(4M)**

Ans. Refer Section 1.6.

3. State the status of recycling of solid waste in India. **(4M)**

Ans. Refer Section 1.7.1.3.

4. Define: (i) Biomedical waste, (ii) E-waste. **(4M)**

Ans. Refer Sections (i) 1.2 (f) and (ii) 1.2 (g).

5. Explain chemical characteristics of solid waste. **(4M)**

Ans. Refer Section 1.5.2.

6. Enlist sources of solid waste. **(4M)**

Ans. Refer Section 1.3.

7. Explain solid waste management techniques hierarchy. **(4M)**

Ans. Refer Section 1.7.1.

8. Explain importance of solid waste management in making any city as smart city. **(4M)**

Ans. Refer Section 1.10.

Summer 2019

1. Enlist eight types of Solid Waste. **(4M)**

Ans. Refer Section 1.2.

2. State the various factors affecting solid waste generation. **(4M)**

Ans. Refer Section 1.8.

3. State two physical characteristics and two chemical characteristics of solid waste. **(4M)**

Ans. Refer Sections 1.5.1 and 1.5.2.

4. Define solid waste. What is its impact on environment. **(4M)**

Ans. Refer Sections 1.1 and 1.6.

5. Explain four R's in waste hierarchy. **(4M)**

Ans. Refer Section 1.7.1.

6. Enumerate various sources of solid waste. **(4M)**

Ans. Refer Section 1.3.

7. Define E-waste. State the composition of E-waste. **(4M)**

Ans. Refer Sections 1.2 (g) and 5.5.2.

8. Describe the hazardous and non-hazardous substances, their origin and their health effects. **(4M)**

Ans. Refer Table 1.1 and 1.2.

9. Describe 'Waste Minimization Approach'. **(4M)**

Ans. Refer Section 1.9.

10. Define recycling. What are the purpose of recycling? **(4M)**

Ans. Refer Section 1.7.1.3.

***

Chapter **2...**

Storage, Collection and Transportation of Municipal Solid Waste

Weightage of Marks = 14, Teaching Hours = 10

Syllabus

2.1 Storage of Solid Waste

2.2 Collection Methods of Solid Waste

2.3 Tools and Equipment - Litter Bin, Broom, Shovels, Handcarts, Mechanical Road Sweepers, Community Bin - like Movable and Stationary Bin

2.4 Transportation of Municipal Waste

2.5 Transportation Vehicles with their Capacity Working - Animal Carts, Auto Vehicles, Tractors or Trailers, Trucks, Dumpers, Compactor Vehicles. Transfer Station - Meaning, Necessity, Location

2.6 Role of Rag Pickers and their Utility for Society

2.7 Organization Pattern of Solid Waste Management System, Practices According to Population of the Town or City

About this Chapter

At the end of this chapter, students will be able to:

- Suggest the relevant method of storage of solid waste for the given site conditions with justification.
- Explain the relevant method of collecting the solid waste in the given situation.
- Implement the relevant techniques for management of solid waste in the given area.
- Suggest the relevant transportation system for transporting the municipal solid waste at the given location with justification.
- Propose the organization chart required to manage solid waste for the given village / town / city of your locality.

2.0 OVERVIEW

- **Municipal Solid Waste (S-15):** It is the waste collected from various sources like household, institutions, commercial centers etc. and then taken to the nearest transfer stations, which is afterwards transferred to Municipal Solid Waste Plants by hauling vehicles for further treatment.
- **Storage** of waste at source is the first essential step of Solid Waste Management. Every household, shop and establishment generates solid waste on day to day basis. The waste

should normally be stored at the source of waste generation till collected for its disposal. In India, such a habit has not been formed and in the absence of system of storage of waste at source, the waste is thrown on the streets, treating streets as store of waste. If citizens show such indifference and keep on throwing waste on streets and expect that municipal sweepers should/would clean the city, the cities will never remain clean. Even if local bodies make arrangements to remove all the waste disposed off by the citizens on the street on day to day basis, the city will remain clean only for two to three hours and not beyond till the habit of throwing waste on the streets is not changed. There is, therefore, a need to educate the people to store waste at source, dispose off the waste as per the directions of the local bodies and effectively participate in the activities of the local bodies to keep the cities clean.

- **Collection** of waste is the second essential step of Solid Waste Management activity. Collection system is necessary to ensure that waste stored at source is collected regularly and it is not disposed off on the streets, drains, water bodies etc. However, step has to co-ordinate well with the first step i.e. Storage of Waste at source.
- **Transportation:** All the waste collected through Collection System, from the households, shops and establishments has to be taken to the processing or disposal site either directly necessitating a large fleet of vehicles and manpower or through cost effective systems which are designed to ensure that all the waste collected from the sources of waste generation is temporarily stored at a common place called "Waste Storage Depots" and then transported in bulk to the processing or disposal sites. Such temporary arrangement for storage of waste is popularly known as dust bin, dhalavs etc. This facility has to be so designed that the system matches with the system of collection as well as transportation of waste.
- Transportation of the waste stored at waste storage depots at regular intervals is essential to ensure that no garbage bins/containers overflow and waste is not seen littered/cluttered on streets. Hygienic conditions can be maintained in cities/towns only if regular clearance of waste from temporary waste storage depots (bins) is ensured.
- Transportation system has to be so designed that it is efficient, yet cost effective. The system should synchronize with the system of waste storage depot and should be easily maintainable.

2.1 STORAGE OF MUNICIPAL WASTE (W-16)

2.1.1 Present Scenario

- Generally no bins for storage of domestic, trade or institutional waste are kept at source. Very few people keep personal bins for storage of domestic, trade or institutional waste at source. The percentage of such people is insignificant.
- Under the situation, most of the domestic waste as well as waste from shops, offices and establishments including hospitals, nursing homes, hotels, restaurants, construction and demolition wastes etc., come on the streets or is disposed off unauthorised on public or private open plots or even discharged in the drains or water bodies nearby resulting in clogging of drains, pollution of water resources and increase in insanitary conditions in the urban areas.
- Hence, there is no practice of storing the waste at source in a scientifically segregated way. Citizens have not been educated to keep domestic, trade and institutional bins for storage of waste at source and stop littering on the streets.

Various Methods of Storage Of Municipal Solid Waste : (S-16)

The methods of storage of solid municipal waste depends upon method of collection of waste. Some of them are given below:

1. Large number of open communal storage sites.
2. Plastic buckets (with lids), with capacities between 7 and 10 lit.
3. Plastic bins (with lids) and equipped with handles, with capacities between 30 and 60 lit.

4. Galvanized steel or plastic bins (with lids), with a capacity between 50 and 70 lit.
5. Disposable plastic bags.
6. Other items commonly used for the storage of wastes include cardboard boxes, kerosene cans and containers made out of truck tyres.
7. **Depots:** A depot typically consists of a single-storey building about the size of a large garage.
8. **Enclosures:** An enclosure is probably the most common communal storage method. Enclosures can have capacities from 1 to 10 m^3.
9. **Fixed storage bins:** These type of containers are usually built from concrete blocks.

2.1.2 Storage of Household Waste

All households may be directed that:

- They shall not throw any solid waste in their neighbourhood, on the street, open spaces and vacant plots or into drains.
- They shall (a) keep the food waste / bio-degradable as and when generated, in any type of domestic waste container, preferably with a cover and (b) keep dry / recyclables wastes preferably in bags or sacks.
- Use of a non-corrosive container with lid is advised for the storage of food/biodegradable/wet waste. A container of 15 litre (0.015 cu.mtr) capacity for a family of 5 members would ordinarily be adequate. However, a household may keep larger containers or more than one container to store the waste produced in 24 hours having a spare capacity of 100% to meet unforeseen delay in clearance or unforeseen extra loads. Wet wastes should preferably not be disposed off in plastic carry bags.
- Keep domestic hazardous waste separately, for disposal at the place may be as arranged for by the Urban Local Bodies (ULB).
- A private society, association of flats/multi-storeyed buildings etc. shall provide a community bin i.e. a bin large enough to hold the waste generated by the members of their society/association for storage of wet domestic wastes and instruct all residents to deposit their domestic waste in this community bin to facilitate collection of such waste by the local body from the designated spot.
- In case of Multi-storeyed buildings where it may be difficult for the waste collector to collect recyclable waste from the doorstep, the association of such buildings may optionally keep one more community bin for storage of recyclable material.
- In slums, where because of lack of access or due to narrow lanes, it is not found convenient to introduce house-to-house collection system, community bins of suitable sizes ranging from 40 to 100 litre (0.04 to 0.1 cu.mtr.) capacity may be placed at suitable locations by the local body to facilitate the storage of waste generated by them. They may be directed to put their waste into community bins before the hour of clearance each day.
- In a situation where local bodies find it difficult to place smaller community bins in slums on account of lack of awareness among slum dwellers, the local bodies may provide larger containers which may match with the local body's transportation system at locations which may be suitable to slum dwellers and convenient for local bodies to collect such waste. Slum dwellers may be directed to deposit their waste in such larger bins before the hour of clearance of waste each day.

2.1.3 Storage of Waste from Shops/Offices/Institutions/ Workshops etc.

All shops and establishments may be directed that:

- They shall refrain from throwing their solid waste/sweepings etc. on the footpaths, streets and open spaces.

- They shall keep their waste on-site as and when generated in suitable containers until the time of doorstep collection.
- The size of the container should be adequate to hold the waste, they normally generate in 24 hours with 100% spare capacity to meet unforeseen delay in clearance or unanticipated extra loads.
- They shall keep hazardous waste separately as and when produced and disposed off as per directions given by the local ULB.
- The association of private commercial complexes/multi-storeyed buildings shall provide suitable liftable community bins which match with the waste collection and transportation system of the local body for the storage of waste by their members and direct them to transfer their waste into the community bin before the prescribed time on a day-to-day basis.
- The association should consult the local body in this matter in advance and finalize the type of bin and the location where such community bin/s shall be placed to facilitate easy collection of such waste.

2.1.4 Storage of Waste from Hotels and Restaurants

All hotels and restaurants may be directed that:

- They shall refrain from throwing their dry and wet solid waste/sweepings on the footpath, streets, open spaces or drains.
- They shall also refrain from disposal of their waste into municipal street bins or containers.
- They shall store their waste on-site in sturdy containers of not more than 100 litre (0.1 cu.m) capacity. The container should have appropriate handle or handles on the top or side and rim at the bottom for ease of emptying.
- In case of large hotels and restaurants where it may not be convenient to store waste in 100 litre or smaller size containers, they may keep larger containers which match with the primary collection and transportation system that may be introduced in the city by the urban local body, to avoid double handling of waste.
- They may be directed to keep hazardous waste separately as and when produced and dispose it off as per the directions of the urban local body.

2.1.5 Storage of Vegetable/Fruit Markets Waste

These markets produce large volumes of solid waste and local bodies may:

- Direct the association of the market to provide large size containers which match with the transportation system of the local body or
- Depending on the size of the market, local body itself may provide large size containers with lid or skips as illustrated below for storage of market waste at suitable locations within markets on full cost/partial cost recovery as deemed appropriate.

2.1.6 Storage of Meat and Fish Markets Waste

- The shopkeepers shall not throw any waste in front of their shops or on the streets or open spaces.
- They shall keep within their premises sturdy containers (of size not exceeding 100 litres i.e. 0.1 cu.m) having lid, handle on the top or on the sides and rim at the bottom of the container with adequate spare capacity to handle expected loads. However, slaughter house wastes should be handled as per the guidelines given by ULB on slaughterhouse waste.

2.1.7 Storage of Street Food Vendors Waste

- All street food vendors may be directed not to throw any waste on streets and pavements.
- They must keep bins or bags for storage of waste that they generate through their activity.
- Their handcarts must have a shelf or bag below for storage of waste generated in the course of business.

2.1.8 Storage of Marriage Halls/Kalyan Mandaps/Community Halls etc. Waste

- A lot of waste is generated when marriage or social functions are performed at these places and unhygienic conditions are created.
- Suitable containers with lids which may match with the primary collection or transportation system of local bodies should be provided by these establishments at their cost and the sites of their placement should be finalised in consultation with urban local bodies to facilitate easy collection of waste.
- On-site bio-digesters for food waste should be encouraged.

2.1.9 Storage of Hospitals / Nursing Homes / Pathological Laboratories / Health Care Centres / Establishments etc. Waste

These establishments produce bio-medical as well as ordinary waste. These may be directed that:

- They shall refrain from throwing any bio-medical waste on the streets or open spaces, as well as into municipal dust bins or domestic waste collection sites.
- They shall also refrain from throwing any ordinary solid waste on footpaths, streets or open spaces.
- They are required to store waste in colour-coded bins or bags as per the directions of the Govt. of India, Ministry of Environment Bio-medical Waste (Management and Handling) Rules 1998 and follow the directions of Central Pollution Control Boards and State Pollution Control Boards from time to time for the storage of biomedical waste.

2.1.10 Storage of Construction and Demolition Wastes (S-15)

- No person shall dispose off construction waste or debris on the streets, public spaces, footpaths or pavements.
- Till finally removed, construction waste shall be stored only within the premises of buildings, or in containers where such facility of renting out containers is available. In exceptional cases where storage of construction waste within the premises is not possible, such waste producers shall take prior permission of the local authority or the State Government as may be applicable for temporary storage of such waste and having obtained and paid for such permission, may store such waste in such a way that it does not hamper the traffic, the waste does not get spread on the road and does not block surface drains or storm water drains.
- Local bodies above 10 lac population must make efforts to provide or encourage the facility of skips/containers on rent for storage and transportation of construction and demolition waste.
- To facilitate the collection of small quantities of construction and demolition waste generated in a city, suitable sites may be identified in various parts of the city and people notified to deposit small quantities of construction and demolition waste. Containers could be provided at such locations and small collection charge levied for receiving such waste at such sites and for its onward transportation. Rates may be prescribed for such collection by local bodies.

2.1.11 Storage of Garden Waste (S-15)

- Private gardens should as far as possible compost and re-use all plant wastes on-site. Where it is not possible to dispose off garden waste within the premises and the waste is required to be disposed off outside the premises, it shall be stored in large bags or bins on-site and transferred into a municipal system on a weekly basis on payment. The generation of such waste should as far as practicable be regulated in such a way that it is generated only a day prior to the date of collection of such waste. It should be stored in the premises and kept ready for handing over to the municipal authorities or the agency that may be assigned for the work of collection of such waste.

- Garden waste and fallen leaves from avenue trees within large public parks and gardens should be composted to the extent possible. However, if such waste has to be disposed off, large containers may be kept, which match with the municipal transportation system for transportation of such waste. Such containers may be provided by local bodies or State Governments owning such parks and gardens. In case of private parks and gardens they should make their own storage arrangement which matches with the municipal primary collection and transportation system.

2.1.12 Storage of Recyclable Waste

- It is essential to save the recyclable waste material from going to the waste processing and disposal sites and using up landfill space. Profitable use of such material could be made by salvaging it at source for recycling. This will save national resources and also save the cost and efforts to dispose off such wastes.
- This can be done by forming a habit of keeping recyclable waste material separate from food wastes, in a separate bag or a bin at the source of waste generation.
- This recyclable waste can be handed over to the waste collectors (rag pickers) at the doorstep.

2.1.13 Parameters to be taken into Account while Making Storage Facility for Solid Waste (W-18)

Following parameters to be taken into account while making storage facility for solid waste.

(i) Efficiency, i.e., the containers should help maximize the overall collection efficiency.

(ii) Convenience, i.e., the containers must be easily manageable both for residents and collection crew.

(iii) Compatibility, i.e., the containers must be compatible with collection equipment.

(iv) Public health and safety, i.e., the containers should be securely covered and stored.

(v) Ownership, i.e., the municipal ownership must guarantee compatibility with collection equipment.

2.1.14 Factors Affecting Generation of Municipal Solid Waste (S-15)

Factors Affecting Generation of MSW:

1. Living standard.
2. Rate of generation of waste.
3. Population growth.
4. Improper guidance.
5. Lack of awareness.
6. Industrialization growth etc.

2.1.15 Measures to be Taken by the Local Bodies Towards Segregation of Recyclable Waste

- Local bodies may mobilise voluntary organisations, Non-Governmental Organisations (NGOs) or co-operatives to take up the work of organising street rag-pickers and elevate them to door step "waste collectors" by motivating them to stop picking up soiled and contaminated solid waste from streets, bins or disposal sites and instead improve their lot by collecting recyclable clean materials from the doorstep at regular intervals of time. Local bodies may, considering the important role of rag pickers in reducing the waste and the cost of transportation of such waste, even consider extending financial help to NGOs and co-operatives in providing some tools and equipment to the rag pickers for efficient performance of their work in this informal sector.
- Local bodies may actively associate resident associations, Trade and Industry associations, Community Based Organisations (CBOs) and NGOs in creating awareness among the people to segregate recyclable material at source and hand it over to a designated waste collector identified by NGOs.

- The local body may give priority to the source segregation of recyclable wastes by shops and establishments and later concentrate on segregation at the household level.
- The upgraded rag pickers on becoming doorstep waste-collectors, may be given an identity card by the NGOs organising them so that they may have acceptability in society. The local body may notify such an arrangement made by the NGOs and advise the people to co-operate.
- This arrangement could be made on "no payment on either side basis" or people may, negotiate payment to such waste collectors for the door step service provided to sustain their efforts.

2.1.16 Provisions of Litterbins on Streets, Public Places etc.

- With a view to ensure that streets and public places are not littered with waste materials such as used cans, cartons of soft drinks, used bus tickets, wrappers of chocolates or empty cigarette cases and the like generated while on a move, litter bins may be provided on important streets, markets, public places, tourist spots, bus and railway stations, large commercial complexes etc. at a distance ranging from 25 metres to 250 metres depending on the local condition. Similar bins for disposal of animal droppings could be placed in posh areas.
- Removal of waste from these litterbins should be done by beat sweepers during their street cleaning operations. Waste from the litterbins should be directly transferred into the handcarts of the sweepers.
- Such facilities of litterbins can be created at no cost to local bodies by involving the private sector and giving them advertisement rights on the bins for a specified period or by allowing them to put their names on the bins as a sponsor. Litterbins should be put in posh as well as poor areas in the proportion decided by local bodies.

2.1.17 Provisions of Special Containers for Storage of Domestic Hazardous and Toxic Waste

- Urban local bodies should provide at strategic locations in different parts of the cities, specially designed containers having two compartments, one for domestic hazardous-toxic waste such as chemicals, batteries etc. and another for sanitary and health related wastes.
- A different colour scheme should be adopted for the containers placed for the storage of these domestic hazardous-toxic wastes. Citizens should be directed to deposit such wastes in such containers only as and when such waste is to be disposed off. Local bodies should insist that such wastes should not be mixed with organic or municipal general waste.

Table 2.1: Summary of Storage of Waste at Source

Sr. No.	Source of Waste Generation	Action to be taken
1.	Households	1. Not to throw any solid waste in the neighbourhood, on the streets, open spaces and vacant lands, into the drains or water bodies. 2. Keep food waste/biodegradable waste in a non-corrosive container with a cover (lid). 3. Keep, dry/recyclable waste in a bin/bag or a sack. 4. Keep domestic hazardous waste if and when generated separately for disposal at specially notified locations.
2.	Multi-storyed buildings, commercial complexes, private societies etc.	1 to 4 as above. 5. Provide separate community bin/bins large enough to hold food/biodegradable waste and recyclable waste generated in the building/ society.

Sr. No.	Source of Waste Generation	Action to be taken
		6. Direct the members of the association/society to deposit their waste in community bin on day to day basis before the hour of clearance.
3.	Slums	1 to 4 as above. 5. Use community bins provided by local body for deposition of food and biodegradable waste.
4.	Shops, offices, institutions, etc.	1 to 4 as above. 5. If situated in a commercial complex, deposit the waste so stored as per 2 and 3 above, in community bins provided by the association.
5.	Hotels and Restaurants	1 to 4 as above. However, the container used should be strong, not more than 100 litre in size, should have a handle on the top or handles on the sides and a rim at the bottom for easy handling.
6.	Vegetable and Fruit Markets	1. Provide large containers, which match with transportation system of the local body. 2. Shop keepers not to dispose off the waste in front of their shops or open spaces. 3. Deposit the waste as and when generated into the large container placed in the market.
7.	Meat and Fish Markets	1. Not to throw any waste in front of their shops or open spaces around. 2. Keep a non-corrosive container/containers not exceeding 100 litre capacity with lid, handle and the rim at the bottom and deposit the waste in the said containers as and when generated. 3. Transfer the contents of this container into a large container provided by the association of the market or local body on day to day basis before the hour of clearance.
8.	Street Food Vendors	1. Not to throw any waste on the street, pavement or open spaces. 2. Keep bin or bag for the storage of waste that generates during street vending activity. 3. Preferably have an arrangement to affix the bin or bag with the hand-cart used for vending.
9.	Marriage Halls, Community Halls, Kalyan Mandaps etc.	1. Not to throw any solid waste in their neighbourhood, on the streets, open spaces and vacant lands, into the drains or water bodies. 2. Provide a large container with lid which may match with the transportation system of the local body and deposit all the waste generated in the premises in such containers.
10.	Hospitals, Nursing Homes etc.	1. Not to throw any solid waste in their neighbourhood, on the streets, open spaces and vacant lands, into the drains or water bodies.

Sr. No.	Source of Waste Generation	Action to be taken
		2. Not to dispose off the biomedical waste in the municipal dust bins or other waste collection or storage site meant for municipal solid waste. 3. Store the waste as per the directions contained in the Government of India, Ministry of Environment Biomedical Waste (Management and Handling) Rules 1998.
11.	Construction and Demolition Waste	1. Not to deposit construction waste or debris on the streets, foot paths, pavements, open spaces, water bodies etc. 2. Store the waste within the premises or with permission of the authorities just outside the premises without obstructing the traffic preferably in a container if available through the local body or private contractors.
12.	Garden Waste	1. Compost the waste within the garden, if possible. 2. Trim the garden waste once in a week on the days notified by the local body. 3. Store the waste into large bags or bins for handing over to the municipal authorities or contractors appointed for the purpose on the day of collection notified.

2.1.18 Effective Storage of Solid Waste (W-15)

- All householders must only present waste for collection by a permitted waste collector.
- All householders must present waste for collection if they are on a collection route or retain all records i.e. receipts/dockets as proof of the proper disposal of their household waste.
- All waste must be stored and presented in appropriate receptacles in a manner that would not endanger health, create a risk to traffic, harm the environment or create a nuisance such as odour or litter.
- All recyclables must be separated at source and presented for collection in appropriate receptacle.
- If a separate organic collection service is made available, this waste shall also be separated at source for collection.
- Receptacles must not be overfilled and lids must be closed at all times.
- Householders must not present waste for collection prior to 6 pm on the eve of collection day.
- Householder must remove any waste containers and/or uncollected waste from point of collection within 12 hrs. of the designated collection time.
- It is advised for the storage of bio-degradable or wet waste that the container should be of capacity 15 lit.

2.2 COLLECTION OF MUNICIPAL WASTE

2.2.1 Present Scenario (S-15)

- Doorstep collection of waste from households, shops and establishments is not very much prevalent in India and wherever it is introduced through private sweepers or departmentally, the system does not synchronize further with the facility of Waste Storage Depots and Transportation of Waste. The waste so stored is deposited on the streets or on the ground outside the dustbin. Thus, streets are generally treated as storehouses of waste and the collection of waste is done, by and large, through street sweeping.

- An appropriate system of collection of waste is to be so designed by the urban local bodies that it synchronizes with storage of waste at source as well as waste storage depots facility ensuring that the waste once collected reaches the processing or disposal site through a containerized system.
- No bins for storage of domestic, trade or institutional waste are kept at source. Very few people keep personal bins for storage of waste. The percentage of these people is very insignificant.
- In most of the situations, domestic waste as well as waste from shops, offices and establishment including hospitals, nursing homes, hotels, restaurants, garden etc. come on the street or is disposed off in open plot or even discharged in the drains. Its result is clogging of drains, pollution of water resources and increasing insanitary condition in urban areas.
- Hence, there is no practice of storing the waste at sources in scientifically segrated way.
- Citizens have not been educated to keep domestic, trade and institutional bins for storage of waste at source.

2.2.2 Various Methods of Collection of Municipal Solid Waste (W-16, 18)

(I) Classification bsed on mode of operation:

1. **Stationary Container System:**
   - In this system, containers used for the storage of waste remain at the point of collection.
   - The collection vehicles generally stop alongside the storage containers, and collection crews load the waste from the storage containers into the collection vehicles and then transport the waste to the processing, transfer or disposal site.
2. **Hauled Container System:**
   - An empty storage container (Known as a drop-off box) is hauled to the storage site to replace the container that is full of waste, which is then hauled to the processing point, transfer station or disposal site.

(II) Classification based on the availability of service:

1. **Alley Service:**
   The containers are placed at the alley line, which are picked up by workmen from refuse vehicle who keep back the empty container at the place.
2. **Set out Set Back Service:**
   Set out man go to the house, collect containers and empty them in the refuse vehicle. Another group of persons return them to the house owner's yard.
3. **Curb (Kerb-side):**
   - A service provided to households, typically in urban and suburban areas for removing household waste is called curb (kerb-side) collection.
   - This is the quickest economical system.
   - 1 driver and 1 or 2 collectors are in crew.
   - In this system, house owner is responsible for placing solid waste containers at the curb on scheduled day.
   - The work man come, collect and empty the container and put back at the curb.
   - House owner is required to take back the empty containers from the curb to the house.
   - There is no need to enter any one's property so no disturbance to house owner.
4. **Backyard Service:**
   - The workers with the vehicles carry a bin, wheel-barrow or sack or cloth to the yard and empty the solid waste container in it. It is then taken to solid waste vehicle where it is emptied.

2.2.3 Measures to Improve the Collection Services (S-19)

- Local bodies should provide daily waste collection service to all households, shops and establishments for the collection of putrescible organic waste from the doorstep because of the hot climatic conditions in our country. This service must be regular and reliable. Recyclable material can be collected at longer regular intervals as may be convenient to the waste producer and the waste collector, as this waste does not normally decay and need not be collected daily. Domestic hazardous waste is produced occasionally. Such waste need not be collected from the doorstep. People could be advised or directed to deposit such waste in special bins kept in the city for disposal.
- Urban local bodies may arrange for the collection of domestic, trade and institutional food/biodegradable waste from the doorstep or from the community bin on a daily basis.
- Local bodies may also arrange through NGOs collection of recyclable waste material/non bio-degradable waste other than toxic and hazardous waste from the source of waste generation at the frequency and in the manner, told by local bodies from time to time in consultation with the NGOs/Resident Associations etc.
- Domestic hazardous/ toxic waste material deposited by the waste producers in special bins (provided by the local body at various places in the city) may be collected at regular intervals after determining the quantities of such waste deposited in special bins.

Measures to be taken to improve the collection services of municipal waste:

1. Awareness regarding segregation of degradable and non-degradable solid waste.
2. Distribution of free colour coded bins to consumers for solid waste storage.
3. Timely and regular service for door to door collection of waste.
4. Participation of rag-pickers in waste collection.
5. Separate bins of larger size for commercial shops and hotels etc.
6. Separate collection system for waste food from marriage halls, hotels etc.

2.2.4 Arrangements to be Made for the Efficient Collection

One or more methods as mentioned below can be used for collection of waste:

- Doorstep collection of waste through containerized handcarts/tricycles or other similar means with active community participation.
- Doorstep collection of waste through motorised vehicles having non-conventional/sounding horns deployed for doorstep waste collection with active community participation.
- Collection through community bins from private societies, multi-storeyed buildings, commercial complexes.
- Doorstep or lane-wise collection of waste from authorised/unauthorised slums or collection from the community bins to be provided in the slums by local bodies.
- House-to-house collection of waste from posh residential areas on full-cost recovery basis.

Sorting at Source: (S-15)

- It is the most efficient technique that reduces the mixing of various biodegradable and non-biodegradable wastes together. In this process, different types of wastes are stored in different containers like plastic waste, biodegradable waste, E-waste, Glass, Paper etc. Based on the nature of waste collected, appropriate treatment is given, which automatically reduces the time and cost of treatment.

2.2.5 Methods of Collection of Waste (S-16; W-16)

2.2.5.1 Door Step Collection Through Containerized Handcarts

- A bell may be affixed to the handcart given to the sweeper or a whistle may be provided to the sweeper in place of a bell.

- Each sweeper may be given a fixed area or beat for sweeping, plus a fixed number or stretch of houses for collection of waste. The local bodies may, based on local conditions, fix the work norms as they think appropriate.
- It is suggested that in congested or thickly populated areas, 250 to 350 running metres (RMT) of road length and the adjoining houses may be given to each sweeper, whereas in less congested areas 400 to 600 running metres of the road length with adjoining houses may be allotted to a sweeper depending upon the density of population in the given area and local conditions.
- In low density areas even 650 to 750 running metres of road length and houses can be given. Normally, 150 to 250 houses coupled with the above road length may be taken as a standard for allotment of work to an individual sweeper.

2.2.5.2 Duties/Role of Sweeper

- The sweeper should ring the bell or blow the whistle indicating his arrival at the place of his work and start sweeping the street.
- The people may be directed through adequate publicity campaign that on hearing the bell or whistle they should deposit their domestic biodegradable waste into the handcart of the sweeper or hand over the waste to him/her.
- At places where it is not convenient for the householder to deposit the waste in the handcart/tricycle, on account of their non-availability at home when sweeper arrives in their areas, they may leave the domestic waste in domestic bins or bags just outside their houses on the street in the morning so as to enable the sweepers to pick up the waste and put it into the handcart.
- No sweeper may be expected or directed to do house-to-house collection by asking for waste at the doorsteps, as this will affect his energy and productivity.

2.2.5.3 Collection by Motorised Vehicles

- Local bodies as an alternative method to doorstep collection by containerised handcarts may deploy motorised vehicles having sounding horn for doorstep collection of waste.
- Driver of the vehicle should intermittently blow the horn announcing his arrival in different residential localities and on hearing this; the householders should deposit their domestic waste directly into such vehicle without loss of time.

2.2.5.4 Collection of Waste from Societies/Complexes

- In private societies, complexes and multi-storeyed buildings, normally no sweepers are provided by local bodies; hence private sweepers are generally engaged. It may therefore be made compulsory for the management of the societies, complexes and multi-storeyed builders, to keep community bins or containers in which dry and wet waste may separately be stored by their residents.
- Such bins may be placed at easily approachable locations to facilitate convenient collection by the municipal staff or the contractors engaged by the local body. The local body should arrange to collect waste from these community bins/ containers through handcarts, tricycles, pick-up vans, or other waste collection vehicles on a daily basis.
- To facilitate collection of waste from societies or commercial complexes, the local bodies should by a rule, make it compulsory for them to identify an appropriate site within their premises for keeping such bin/container for the storage of waste.

2.2.5.5 Collection of Waste from Slums

- Local bodies should collect waste from slums by bell ringing/whistle system along their main access-lanes. Residents should bring their wastes from their houses to hand carts.

- Where slum residents prefer community bins, they should bring their biodegradable waste to these bins only an hour or two before the time of clearance.
- The local body may, if so desired, engage a private contractor for collection of such waste. Performance certification by a "Mohalla (local level) Committee" may be insisted upon in such cases.

2.2.5.6 Collection at the Doorstep in Posh Residential Areas

- In posh residential areas where the residents as a whole might not be willing to bring their waste to the municipal handcart/tricycle, system of collection from the door step on full cost recovery basis may be introduced.
- This service can be contracted out by the local body or NGOs or contractors registered with the local body may be encouraged to provide such service in the areas where it is found economically viable to introduce door to door waste collection service.
- This service may not be provided to isolated houses, shops and establishments.
- Penalty provisions may be introduced for failure of service where contracts are proposed to be awarded.

2.2.5.7 Collection of Dairy and Cattle-Shed Waste

- The dairies and cattle breeders having sheds within the city limits should be asked to move the cattle sheds outside the city limits and until this is implemented, they should be directed not to stack the cow dung, grass or other stable wastes within their premises or on the roadside.
- They must transfer the waste produced by them daily into the specified municipal storage containers nearby, which should be collected at regular intervals by local bodies.

2.2.5.8 Collection of Domestic Hazardous and Toxic Waste

- Collection of domestic hazardous and toxic waste such as used batteries, paints, broken tube lights, expired medicines and others kept separately in the bins placed in various parts of the city should be collected periodically by the urban local bodies and got segregated and disposed off as per the hazardous waste management rules of the Government of India.

2.2.5.9 Collection of Duly Segregated Recyclable/Non-bio-degradable Waste from Households (S-19)

- Recyclable waste has a value. Several rag pickers in the urban areas, therefore, move from street to street, bin to bin and go to the dump yard to pick up recyclable waste.
- These rag pickers are exposed to health risks as they put their bare hands in contaminated waste. They sell contaminated waste to the waste purchasers stored in slums creating unhygienic conditions. Quite often they spread the waste at the dust bin site to pick up recyclable waste.
- This system can be improved by introducing a system of collecting recyclable waste from the doorsteps changing the roll of rag pickers to that of waste collectors. This informal sector could thus be organised through NGOs, upgraded and given an opportunity to earn their living through doorstep collection of unsoiled recyclable waste.
- NGOs may be activated to organise the rag pickers and convert them into door-step waste-collectors to improve their quality of life and to reduce their health risk. This will also increase their income levels.
- NGOs may allot to such waste collectors specified lanes and by-lanes comprising of 150 to 250 houses and some shops for doorstep collection of recyclable. They may also be given identity cards by the NGOs for increasing their acceptability in society.
- NGOs and/or the corporation may support such waste collectors by giving them bags and tools required for collection of recyclable waste from the doorsteps.

Waste Segregation and Its Importance: (S-19)

Waste segregation means separation of waste at source so that it can reduce following problems –

- Waste segregation is included because it is much easier to recycle. Effective segregation of wastes means that less waste goes to landfill, which makes it cheaper and better for people and the environment. It is also important to segregate for public health. In particular, hazardous wastes can cause long-term health problems, so it is very important that they are disposed off correctly and safely and not mixed in with the normal waste coming out of your home or office.
- If proper waste segregation is not done, then following bad effects will happens:
    1. For public, health problems arise from the breeding of diseases due to flies, insects and rats.
    2. The health problems are more serious when transfer of pollution to water, ground water and air takes place.
    3. Air pollution causes due to insufficient burning of wastes in open or in plants due to improper gaseous effluent removal devices.
    4. Mixing of hazardous wastes from industries with municipal wastes create risks to human health.
    5. Danger of concentration of heavy metals in the food chain due to industrial effluent discharges into drainage system creates danger to public health.
    6. Sanitary workers dealing with municipal solid waste are infected with gastrointestinal parasites, worms etc.
    7. The organic fraction of municipal solid waste provides food and shelter to insects and rodents which causes 22 human diseases due to improper solid waste management causing adverse impact upon public health and environmental quality.

Measures to be Taken Towards the Segregation of Recyclable Waste:

1. The local body may mobilize NGO's or co-operatives to take up the work of organizing street rag-pickers and convert them to door-step waste collectors by motivating them.
2. The local bodies may actively associate resident associations, trade and industry associations and NGO's in creating awareness among the people to segregate recyclable material at source and hand it over to a designated identified waste collector.
3. The local body may give priority to the source segregation of recyclable waste by shops and establishments and later concentrate on segregation at the household level.
4. The upgraded rag-pickers on becoming door-step waste-collectors, may be given an identity card by the NGO's organizing them so that they may have acceptability in society.
5. Keep separate large bins for recyclables near the society.
6. Products and packaging materials should be conspicuously labelled to indicate recycled content, including post-consumer content, recyclability, toxicity and appropriate disposal.
7. The establishment of stable markets for recycled materials is essential. Awareness should be created to promote procurement of products containing a high content of recycled and recyclable materials.

2.2.5.10 Collection of Waste from Shops and Establishments (S-19)

- Shops and establishments normally open after 9 am. These timings do not synchronize with the usual work schedule of sweepers. Under such a situation one of the following alternatives may be adopted.

1. Sweepers may first carry out the work of street sweeping in the morning hours as usual and soon thereafter take up the work of door-step collection of waste, after most of the shops have opened.

2. Waste collectors (rag pickers) may be organised to collect the recyclable waste from shops and establishments as soon as they open, as most of such waste is recyclable. Working arrangements may be made with the shops and establishments accordingly. The shops and establishments may be asked to store waste in two bins if they produce waste other than recyclable waste also. This arrangement may be made on 'No payment' basis on either side.
3. The recyclable material received by the waste collectors directly from shops and establishments would give them a better return. The waste would be dry and not soiled and would fetch a good price in the market. This will work as an incentive for them to continue door to door collection.
4. The associations of markets, shops and establishments may be persuaded to organise this service with the help of NGOs and waste collectors in their market.
5. Doorstep collection service from shops and establishment may be provided or may be contracted out on 'full cost-recovery' basis.
6. Large commercial complexes could use 3.0 cu.m to 7.0 cu.m containers, which are commonly used by the local bodies for community storage of waste so that its transportation could be co-ordinated by local bodies along with other containers, kept in the city.

2.2.5.11 Collection of Bio-medical Waste

- Collection of bio-medical waste should be done in accordance with the rules/directions contained in the Ministry of Environment and Forests, Govt. of India Notification dated 20th July 1998, as the liability for safe disposal of biomedical waste is on such waste producer and the local body as such is not directly responsible to provide any service.

2.2.5.12 Collection of Hotel and Restaurant Waste

- Hotels and restaurants may make their own arrangements for collection of waste through their own association, or local bodies may extend help in primary collection of such waste by deploying their own manpower and machinery for door step collection of such waste on full-cost-recovery basis. The cost could be recovered on pro-rata basis. Doorstep service may be contracted out by local bodies if so desired.
- Charges for the collection of hotel waste may depend upon the quantity of waste to be picked up from the hotels and restaurants and frequency of collection required.
- The cost recovery may be planned according to the classification of hotels/ restaurants made on the above basis and decided in consultation with them.
- Thorough survey of the waste generation by hotels/ restaurants may be made before the collection rates are introduced and notified.

2.2.5.13 Collection of Construction and Demolition Waste

- Local bodies should prescribe the rate per tonne for the collection, transportation and disposal of construction waste and debris and notify the same to the people.
- Every person who is likely to produce construction waste may be required to deposit with the concerned local body an approximate amount in advance at the rates as may be prescribed by the local body from time to time, for the removal and disposal of construction waste from his premises by the local body. Such amount may be deposited at the time when the building permission is being sought and in cases where such permission is not required, at any time before such waste is produced.
- The charges for removal of construction waste to be doubled for those who fail to deposit the amount in advance.
- Large local bodies may provide skips (large containers) to the waste producers on rent for the storage of construction waste so that double handling of the waste can be avoided or use front end loader and trucks to pick up such waste. In small towns this may be done manually using trucks, tractors and manpower.

- To facilitate disposal of small quantities of construction/demolition waste, large containers may be placed in various parts of the city where waste producers may deposit small quantities through private labourers, cart pullers, donkeys etc. and such waste may be collected by ULBs from time to time before such containers start overflowing.

2.2.5.14 Collection of Waste from Marriage Halls, Kalyan Mandaps, Community Halls etc.

- The special arrangement should be made for collection of waste from marriage halls, kalyan mandaps, community halls etc. daily on a full-cost recovery basis.
- The cost of such collection could be built into the charges for utilising such halls. This service may be provided preferably through a contractor or departmentally as the local bodies deem fit.
- On-site processing of food wastes by bio-methanation and composting may be encouraged.

2.2.5.15 Collection of Waste from Gardens

- The waste stored in public and private parks, gardens, lawn plots etc. should be collected on a weekly basis by arranging a rotation for collecting such waste from different areas, on different days to be notified to the people to enable them to trim the trees and lawns accordingly and keep the waste ready.
- This waste may be got collected through a contractor or departmentally. Cost recovery may be insisted upon, based on the volume of waste collected.

2.2.5.16 Collection of Waste from Vegetable, Fruits, Meat and Fish Markets

- Such wastes should be removed on a daily basis either departmentally or through contractors on full or part-cost-recovery basis as decided by local bodies.
- Large containers kept in the fruit and vegetable markets should be removed during night time or non-peak hours and the waste from meat and fish markets should be collected through closed pick-up vans service by engaging a contractor or departmentally.

2.2.6 Health Problems during Time of Segregation, Reuse, Recovery and Recycling of Solid Waste (S-17; W-18)

The scavengers or rag-pickers work under extensive health risks, which are largely undocumented, and suffer severe exploitation and deprivation. The possible health hazards to human being include:

1. **Communicable diseases:** Houseflies may be important in the transmission of enteric infections, particularly responsible for infant diarrhoea and dysentery.
2. **Non-Communicable diseases:** Solid waste during its segregation, reuse/recovery/ recycling may cause eye irritation, respiratory diseases, hand/leg injuries, poisoning, hearing defects etc.
3. **Aesthetics:** Solid waste causes odour, visibility and dust problem to human being.

2.3 TOOLS AND EQUIPMENT

Tools and Equipment for Collection and Transportation of a Solid State: (S-15)

1. Litter bin
2. Broom
3. Shovels
4. Handcarts
5. Animal carts
6. Tractors and Trailers
7. Auto vehicle
8. Trucks.
9. Dumper
10. Compactors vehicles.

2.3.1 Litter Bin (W-18)

2.3.1.1 General Description

- The litter bins are to be used for collecting the waste to be thrown by citizens/public moving on the roads/pavements as equipment for primary collection of garbage.

- These shall be placed/grouted at the sides of road/pavements so that person willing to throw waste can use it. This may also be used as a device for primary collection after sweeping a particular lane.
- The litter bin/s shall be good looking. The frame shall be grouted for its rigidness and security. The bins shall be hanged on this frame at suitable height so that public can easily put the waste and it can be easily unloaded by swinging.
- These are normally available in capacities of 60 and 150 litres.

2.3.1.2 General Requirements

- The internal form and surface of the container shall be such that it does not trap the contents.
- There shall be no sharp edges on the container.
- The internal and external surfaces shall be smooth and non-porous, free from cracks, splits, dents, distortion, blisters, voids, air bubbles and other surface blemishes or defects.
- The bins are expected to be used in outdoor conditions. They should be able to withstand outdoor weather conditions in India.
- The test certificate of U.V. stabilization from original manufacturer of polyethylene is to be supplied with bins. The approved make of the polyethylene material shall be IPCL/GAIL/ Reliance.
- The moulded plastic container shall be made in one piece through Injection/ Roto Moulded Process with top rim outside.
- The bins shall conform to the colour code and have embossment of Logo of ULBs and any other markings as specified through a sticker fixed on the body of container.
- The bins shall be provided with adequate stiffeners/ribs on all sides and at bottom for enhancing its structural strength.
- The front rim of the bin shall be fitted with support and guide ribs and bar net of high capacity, double angle edge for greater safety during tipping.
- The bin shall conform to the dimension and other particulars, material specifications and body construction and various tests as prescribed by ULBs.

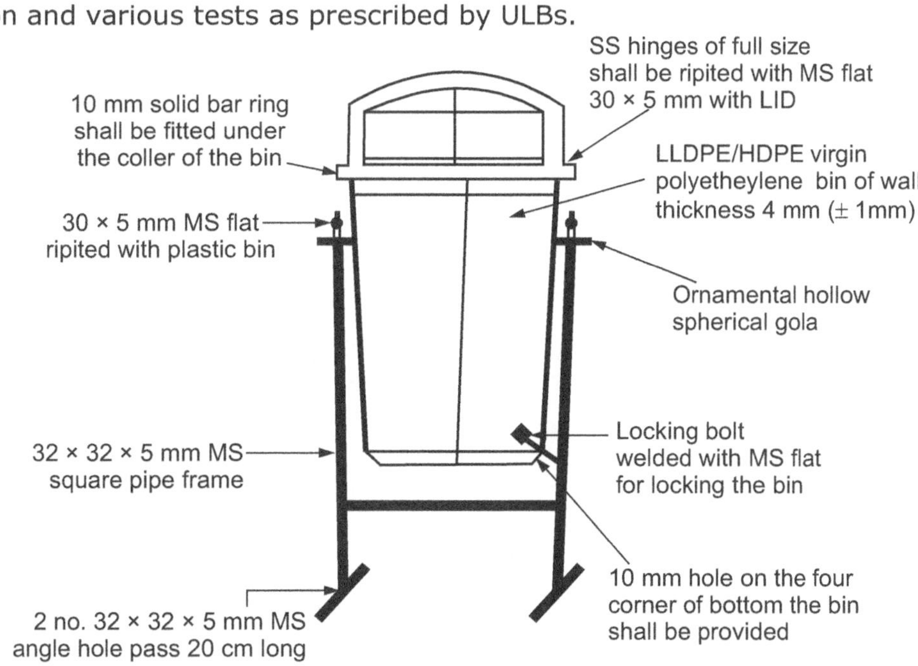

Fig. 2.1: Road Side Litter Bin of 60 Litre Capacity with MS Frame

2.3.2 Broom

- This is used for sweeping the streets.
- Length of broom 80 - 85 cm, weight of broom 1 kg, binding material 20 gauge m sheet ring having width of 1.5 - 2 cm.
- Handle of broom is bamboo of 135 cm length and 3 - 4 cm diameter, weight of the handle is 0.9 kg.

Fig. 2.2: Brooms

2.3.3 Shovels

- A shovel is a tool for digging, lifting and moving bulk materials, such as soil, coal, gravel, snow, sand, garbage or ore. Shovels are common tools that are used extensively in agriculture, construction, lifting the garbage and gardening.
- Most shovels are hand tools consisting of a broad blade fixed to a medium-length handle. Shovel blades are usually made of sheet steel or hard plastics and are very strong. Shovel handles are usually made of wood (especially specific varieties such as ash or maple) or glass-reinforced plastic (fibreglass).

Fig. 2.3: Shovel

2.3.4 Hand Carts

- Hand cart is fabricated out of MS angle, Tee and Flats, suitable to accommodate 6 or 8 bins of 25 litre capacity and ideal for solid waste collection and transfer to central collection collectors.
- The cart shall be strong and sturdy and suitable to satisfy the critical needs of waste collection and transportation.
- The hand cart should be of minimum size 1010 mm × 700 mm × 250 mm with 6 compartments for accommodating 6 numbers of HDPE bins of 25 litres capacity.

Fig. 2.4: Hand Cart

2.3.5 Mechanical Road Sweeper (S-18)

- It is a machine that cleans streets. A street sweeper cleans the streets, usually in an urban area.
- Street sweepers have been employed in cities since sanitation and waste removal became a priority. A street-sweeping person would use a broom and shovel to clean off litter, animal waste and filth that is accumulated on streets. Later, water hoses were used to wash the streets.

- Machines were created in the 19th century to do the job more efficiently. Today, modern street sweepers are mounted on truck bodies and can vacuum debris that accumulates in streets.
- Newer street sweepers are capable of collecting small particles of debris. Many street sweepers produced today are PM10 certified, meaning that they are capable of collecting and holding particulate matter sized less than 10 μm.
- Modern street sweepers are equipped with water tanks and sprayers used to loosen particles and reduce dust. The brooms gather debris into a main collection area from which it is vacuumed and pumped into a collection bin or hopper.
- A regenerative air street sweeper uses forced air to create a swirling effect inside a contained sweeping head and then uses the negative pressure on the suction side to place the road debris inside a hopper. Debris is removed from the air by centrifugal separation and reused, keeping particulate matter inside the hopper.

Fig. 2.5: Mechanical Road Sweeper

Fig. 2.6: Circular Broom under Road Sweeper

2.3.6 Community Bins (Movable and Stationary)

- Community bins are large and meant for use at centrally located pick-up points. Movable bins are provided with wheels and handles to manage unloading and cleaning easily and efficiently.
- Stationary bins are without wheels but have handles and swinging system to unload.
- Their capacities are in the range of 750-1000 litres.

Fig. 2.7: Community Bin

2.3.7 Centralized Sorting of Solid Waste (W-15)

- Centralized sorting of solid waste means separation of bulky items, separation of waste components by size. It is done by using screens or manually.
- Also separation of ferrous and non-ferrous metals is done on magnetic belt.
- Separation of glass and plastic items is also done.

2.4 TRANSPORTATION OF MUNICIPAL WASTE (W-18)

- Transportation of the waste stored at waste storage depots at regular intervals is essential to ensure that no garbage bins/containers overflow and waste is not seen littered on streets.
- Hygienic conditions can be maintained in cities/towns only if regular clearance of waste from temporary waste storage depots (bins) is ensured.
- Transportation system has to be so designed that it is efficient and cost effective. The system should match with the system of waste storage and should be easily maintainable.

2.4.1 The Present Scenario/Problems in Transportation of Waste

2.4.1.1 Inefficient and Unscientific Manual Loading of Waste

- In most of the cities/towns there is no synchronization between waste storage and transportation of waste. Waste stored in open spaces is either loaded manually or with the help of loaders in traditional trucks.
- Manual loading takes time and reduces the productivity of the vehicles and manpower deployed. Besides, manual handling of waste poses a threat to the health of the sanitation workers as the waste is highly contaminated.
- Here the loading operation is cheap but the loaders can not clean waste storage places fully, besides loading machine damage the flooring and screen walls very often necessitating frequent repairs. If repairs are not carried out on time, the damaged flooring becomes a source of nuisance.
- In cities where Dhalavs or large masonry bins are used, waste is not regularly removed from inside of such structures and loaders can not effectively function for removal of waste from the corners of such structures. Leftover (uncollected) waste rots and gives foul smell causing nuisance and insanitary conditions.

2.4.1.2 Issues Regarding Transportation of Solid Waste (S-15)

- Compactors and properly covered dump trucks were to be utilized to transport the waste to the landfill site.
- Separate vehicles were to be employed for transportation of biodegradable waste and mixed recyclable waste.
- Compactor loaders directly lift the bin, unload the waste and replace it in the original position.
- Choice of vehicle depends on the access roads to individual sites.
- Dump trucks are fitted with hydraulic equipments which enable them to unload waste without the help of manual labour.

2.4.1.3 Irregular Transportation

- Cities and towns generally have limited fleet of vehicles and most of them are old and require frequent repairs with the result that the transportation of waste does not take place regularly.
- The waste is generally seen lying in heaps or scattered at the unscientifically designed dust bins giving unsightly appearance besides causing nuisance and unhygienic conditions.

2.4.1.4 Underutilization of Fleet of Vehicles

- Most of the vehicles are manually loaded and the lorries which can easily take 5 to 6 tonnes of solid waste in one trip, carry only 1-3 tonnes of waste as strict monitoring system does not exist.

- In several cities, small vehicles and even bullock carts are taken directly to landfill sites located at long distances. This makes the transportation operation very inefficient and uneconomical.

2.4.1.5 Open Trucks Cause Nuisance

- Open trucks loaded with garbage pass through cities and towns and release foul smell and cause nuisance to people.
- At places where cover material is arranged, covering of trucks is done half-heartedly and nuisance continues.

2.4.1.6 Non-Routing of Vehicles

- In cities and towns, by and large, transportation net-work is ill-designed.
- Waste storage places are not cleared at regular intervals and more or less firefighting operations are carried out by local bodies. These sites are attended to more on the basis of the complaints received or pressure brought on local staff rather than following a system of regular removal of waste from waste storage depots. The system of routing of the vehicles and the clearance of the bins on day to day basis thus generally breaks down.

2.4.1.7 Transportation of Waste from Hotels, Restaurants, Hospitals, Construction Sites Etc.

- In many cities there is no separate system of collection and transportation of such wastes and waste from the above sites is not cleaned regularly.

2.4.2 Measures to be Taken to Improve the System

2.4.2.1 Domestic/Trade/Institutional Waste (W-15)

Transportation of waste from temporary waste storage depots/sites may be planned in accordance with the frequency of containers becoming full. The locations where the containers are placed may be grouped into four categories as under:

- Containers which are required to be cleared more than once a day.
- Containers which are required to be cleared once a day.
- Containers to be cleared on alternate days.
- Containers, which take longer time to fill and need clearance twice a week.

It may be ensured that vegetable, fruits, meat, fish market waste is removed at least once in a day.

2.4.2.2 Routing of Vehicles

- Depending on the containers to be cleared each day, the route for lifting containers may be worked out avoiding haphazard movement of vehicles to the extent possible. This will save a lot of fuel and time.

2.4.2.3 Use of Vehicles in Two Shifts

- All the vehicles may be utilized in two shifts to lift containers, to ensure full utilization of the fleet of vehicles and to reduce the requirement of new vehicles.
- Transportation of waste during night time may be done in areas where there is serious traffic congestion during the day and it hampers solid waste management operations. Work at night will increase the productivity and reduce the cost of such service.

2.4.2.4 Type of Vehicles to be Used

- Vehicles which can synchronize well with containers placed at temporary waste storage depots should be utilized for transportation to prevent multiple handling of waste. The selection of the type of vehicles should also be done keeping in mind the quantity of waste to be transported, the distances to be travelled, the road widths, road conditions, work shop facilities etc.
- In cities above 5 lac population, hydraulic vehicles could be used.

- Following type of vehicles or similar vehicles could be used for transportation of waste:
  - Container lifting devices such as dumper placers/skip lifters or similar other vehicles may be used for transportation of large size containers to transfer stations or to disposal sites.
  - At places where small size containers of 0.5 to 1.0 cu.m. may have been placed, the refuse collector machine without compaction devices of 6 to 15 cu.m. capacity having top or back loading facility may be used. This vehicle, instead of transporting the container, lifts and unloads the contents of the small container into the body of the vehicle through a hydraulic system and puts the empty container back in place.
  - Local bodies should provide facility of direct loading of waste in small vehicles in the congested areas where small vehicles may be parked at suitable locations during primary collection and sweepers may be directed to transfer the waste from the hand carts directly into vehicles. The waste so collected in small vehicles may be taken to the nearest transfer station to transfer the contents into a large container by using a ramp to reduce the cost of transportation.
  - In small cities with poor repairs and maintenance facility, where hi-tech vehicles may not work efficiently, tractor-trolley combination or lifting of containers or toeing of containers by tractors may be utilized. Simple hydraulic tipping-trailers are recommended to avoid manual unloading at processing plants or disposal sites.

2.4.2.5 Bio-Medical Waste from Hospitals, Nursing Homes, Health Care Establishments Etc.

- Transportation of bio-medical waste has to be arranged by waste producers or their associations.
- Adequate protective clothing must be used by the staff handling biomedical waste.

2.4.2.6 Transportation of Waste from Hotels and Restaurants

- Hotels and restaurants waste should be collected once or twice daily through a contract given by the associations of hotels and restaurants, or at their request by local bodies on cost recovery basis.
- Doorstep collection system may be introduced for collection of such waste. Either refuse collector with back loading facility or motor vehicle with close body may be used.
- The entire collection and transport system could be privatized and rates may be prescribed by the associations or local bodies. 33% spare vehicles may be kept in reserve to ensure reliable service.

2.4.2.7 Transportation of Construction Waste and Debris

- Disposal of construction and demolition waste and debris is the liability of waste producers. If such waste is not promptly removed within a reasonable time prescribed by the local body, it may be removed by the local bodies themselves on full-cost-recovery basis.
- One of the following methods may be adopted for transportation of construction waste and debris:
  - In very large cities where a skip-renting system can be introduced, the skips may be transported by hydraulic system at a time mutually agreed upon between the local body and waste producer.
  - When sufficient cost-recovery fee is deposited in advance by waste producer for removal of construction waste, such waste may be loaded mechanically into skips or vehicles using front-end-loaders. One front-end loader and 3 to 4 trucks can transport 100 to 150 tonnes of construction waste in one shift.
  - In small cities under 5 lac population, construction waste may be manually loaded into trucks or tractor trolleys and transferred to disposal sites.
  - Since all such waste must be cleared sooner or later, the more promptly this is done, the cleaner the city will be and there will be less traffic obstruction.

2.4.2.8 Transportation of Waste from Narrow Lanes

- Quite often small quantities of waste are disposed off in narrow lanes, which cannot be removed by sending out the usual transport vehicles.
- Loading rickshaws or traditional carts or animals may be used for removal of such waste manually but very promptly.

2.5 TRANSPORTATION VEHICLES WITH THEIR CAPACITIES AND WORKING

(S-16; W-16)

- The collection of refuse involves all of the steps necessary for transferring the solid wastes from the storage point to the place of treatment or disposal. The process involves emptying the storage container into a vehicle in which the wastes are transported. The collection service may be designed in different ways and can use several transport methods. Transportation methods range from handcarts to 30-ton vehicles.
- The following goals are applicable to transportation vehicles of all types:

1. The load of waste should be thoroughly covered during transport; this is particularly important for motor vehicles travelling at 30 km/hr or more.
2. The loading height of vehicles receiving the contents of containers emptied manually should not exceed 1.6 m.
3. The body of a vehicle should have power-operated or hand-operated tipping gear, or a power-operated ejection plate, unless the load of waste is carried in portable containers.
4. If the vehicles used for primary collection are handcarts or those drawn by an animal, the vehicles should receive the same standards of mechanical design as motor vehicles, including the use of bearings for the wheels and rubber or pneumatic tyres.

Transport Vehicles with Their Capacities: (W-16)

1. Handcart: Consists of 6 or 8 bins of 25 lit capacity.
2. Animal cart: 1.5 m^3.
3. Auto vehicle: 2 m^3.
4. Dumper: 12 m^3.
5. Compactor: 5-10 m^3.
6. Tractor and Trailer: 6 m^3.
7. Truck: 10 m^3.

2.5.1 Animal Carts

- Horses were widely used in North America and in Europe for door-to-door refuse collection until World War II. Horses, mules, donkeys and other animals are used for waste collection in several countries around the world.
- A typical cart is shown in Fig. 2.8 below. The capacity of draught animal carts generally ranges between 2 and 4 m^3. In some cases, the carts are equipped with bodies that can be tipped, by either pivoting the body or by using a manually-operated worm and nut mechanism.
- Animal carts have the following **advantages**: No consumption of fossil fuels, very low capital and operational cost compared with motor vehicles and very quiet operation.
- The low speed of animal carts limits their effective radius of operation to about 3 km. In streets with heavy traffic, animal carts may interfere with motorised traffic.
- Animal carts can and do operate in conjunction with a two-level transfer station in which they tip their loads directly into a large motor vehicle. Some cities have successfully operated this system and have taken advantage of both methods of transport: animal carts for the slow "stop-and-go" portion of the operation and motor vehicles for "high-speed" transport from the collection point to the distant final disposal site.
- However, despite their popularity, more attention needs to be paid to the design of the carts.
- Ideally, animal carts should have bodies fabricated from steel and mounted on pneumatic tyres.
- The carts should be low-loading, fitted with sliding shutters and equipped with manually operated tipping gear.

Fig. 2.8: Animal Cart

2.5.2 Auto Vehicles

- The two-stroke, three-wheel motorcycle is a very common means of transportation in several developing countries and is a viable alternative for waste collection.
- The tricycle can be fitted with a high-level tipping body of about 2 m^3 capacity while retaining a low-loading line.
- The motorised tricycle is in common use in several cities in Asia, particularly in the old sections of the cities where the streets are too narrow to allow passage of larger vehicles. Its relatively high speed which gives this system an operating radius of about 10 km.
- However, the tricycle does not operate well on rough, unpaved roads. If the road system to the disposal site is inadequate, tricycles should discharge at either transfer facilities or processing plants.

2.5.3 Tractors and Trailers

- The agricultural tractor is one of the most utilised motor vehicles for collecting waste in economically developing countries.
- A tractor has several advantages over other types of motorised vehicles.

  Some of these **advantages** include: relatively low capital cost, capacity to transport a large load relative to energy use, readily available maintenance facilities, manoeuvrability on a landfill due to large tyres and high torque and ability to use the power take-off to operate hydraulic tipping gear on a trailer. Despite its relatively low road speed (about 20 km/hr), tractors offer one of the least expensive methods of motor transport of solid wastes, upto a trailer capacity of about 6 m^3.
- There are a number of types of agricultural tractor-trailer systems:

  Mini-agricultural tractors or jeeps can be used with shuttered side-loading trailers upto 4 m^3;

  Full-size agricultural tractors can be used with trailers upto 6 m^3; and Articulated semi-trailers are available with capacities upto 30 ton for long-distance transfer.
- An example of a tractor-trailer system is presented in Fig. 2.9 below.
- The agricultural tractor with trailer often is used as a single unit for the collection of refuse from houses or communal storage points. This combination also has substantial potential as a transfer unit because of the ease with which the tractor can be separated from the trailer.

Fig. 2.9

2.5.4 Trucks

- This type of vehicle is available almost worldwide. It is primarily designed for the transport of construction materials. However, it is also widely used for the collection of wastes from communal sites.
- The body of the truck is usually made of steel, with a flat platform equipped with hinged sides and tail-boards about 40 to 60 cm high. The volume of the truck is usually about 5 to 6 m^3 and is suitable to carry high-density materials such as bricks and aggregates.
- One of the major disadvantages of the vehicle in its standard form is that, it is rarely able to carry its rated payload of solid wastes. Even high-density wastes piled on the vehicle would be unlikely to exceed 4 ton.
- It is, therefore, common practice to extend the height of the sideboards in order to increase the volumetric capacity. This practice, however, makes it necessary to either use ladders to load the vehicle or to place workers inside the body to receive containers handed up to them by collectors.
- The **advantages** of this type of truck are the following: It is relatively inexpensive, it is sturdy and easily obtainable, it has good ground clearance and it performs well on rough roads.
- In applications involving the collection of solid wastes, the truck should have a carrying capacity of at least 2 m^3/ton. In addition, the loading height should not exceed 1.6 m.
- The shutters are usually plain sheets of metal running in grooves. The load can be distributed within the body by the use of rakes. During the final stages of loading, the waste can be piled against closed shutters along one side. Because of the potential difficulty of unloading, it is advisable to equip these vehicles with hydraulic tipping gear.

2.5.5 Dumper (Also known as Fore and Aft Tipper)

- This design appeared in the mid-1930s and was used in Europe until about ten years ago. Its distinguishing feature is that, the body can be tipped in two ways: toward the front of the body during the loading process and toward the rear for unloading.
- This relatively simple mechanism achieves a result similar to the hydraulic ram at the rear of a compactor vehicle. However, the compression effect is much lower than that obtainable with the compaction unit. The forward tipping operation may be required about 12 times per load. A suspended barrier inside the body prevents the wastes from falling back after tipping.

- This type of vehicle utilises body capacities of about 12 m^3.
- This design approximates that of a compactor and is suitable for densities from 250 kg/m^3 and upward. The vehicle can be built on a standard chassis with normal wheel diameter and presents few maintenance problems.
- A schematic diagram of the fore and aft tipper is shown in Fig. 2.10 below.

Fig. 2.10: Dumper

2.5.6 Compactor Vehicles

- The garbage compactor truck usually compacts the waste using a hydraulically powered mechanism that employs a moving plate or shovel to scoop the waste out from the loading hopper and compress it against a moving wall.
- In most compactor designs, the plate has a pointed edge (hence giving it the industry standard name packer blade) which is designed to apply point pressure to the waste to break down bulky items in the hopper before being drawn into the main body of the truck.
- Compactor designs however, have been many and varied. The most popular being the traditional "sweep and slide" system used a combination of a lifting loading hopper and a sweeper blade to clear and compact waste in anticipation of the next load.
- Capacity varies from 5 m^3 to 10 m^3 and have hydraulic system with rear loading arrangement.

Fig. 2.11: Compactor Vehicle

2.5.7 Goals/Requirements Applicable to Transportation Vehicles for Solid Waste (W-15)

- The waste is transported from the storage depots to the disposal sites in tractor trolleys or ill designed open trucks.
- It has been instructed by the law that, the transportation must be done in closed containers only.
- The industrial waste must be transported separately and must be disposed in a safe way after suitable treatment.
- Any type of the hazardous waste should be labelled and coded so that in case of an accident the emergency services know, how to handle a spillage.
- The loading height of vehicles receiving the contents of containers emptied manually should not exceed 1.6 m.
- Transportation system has to be so designed that it is efficient and cost effective.

2.5.8 Transfer Station

2.5.8.1 Meaning (S-15, 16, 18, 19; W-15, 16)

Transfer Station:

- These are the open or closed structures built by competent authority at various locations in city and waste collected by hauling vehicles is initially transferred to these stations.

2.5.8.2 Location **(W-18)**

- A transfer station is a building or processing site for the temporary deposition of waste. Transfer stations are often used as places where local waste collection vehicles will deposit their waste load prior to loading into larger vehicles.
- These larger vehicles will transport the waste to the disposal site either in an incinerator, landfill, or hazardous waste facility, or for recycling.

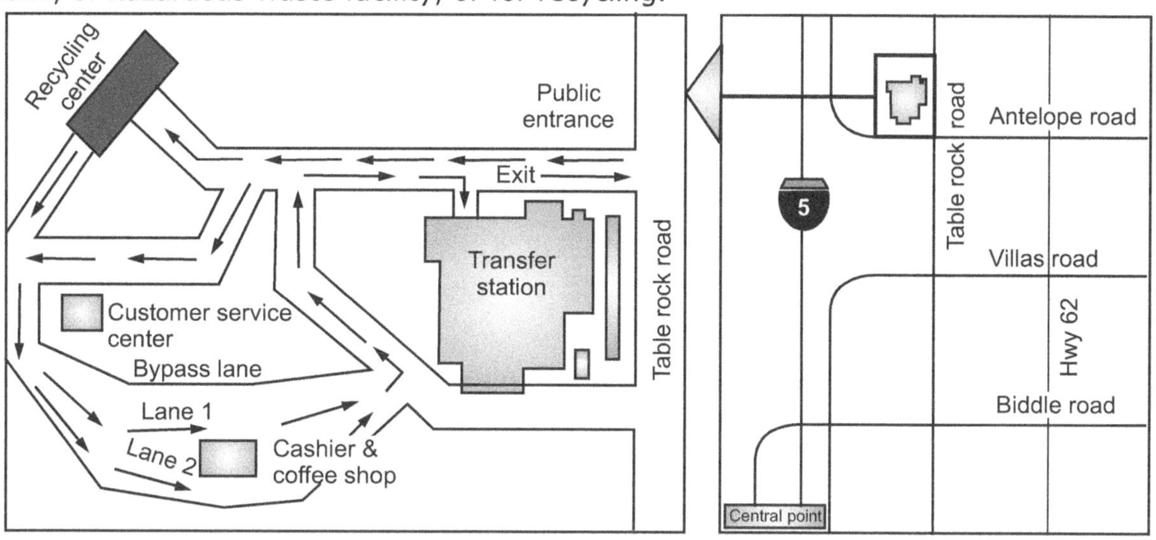

Fig. 2.12

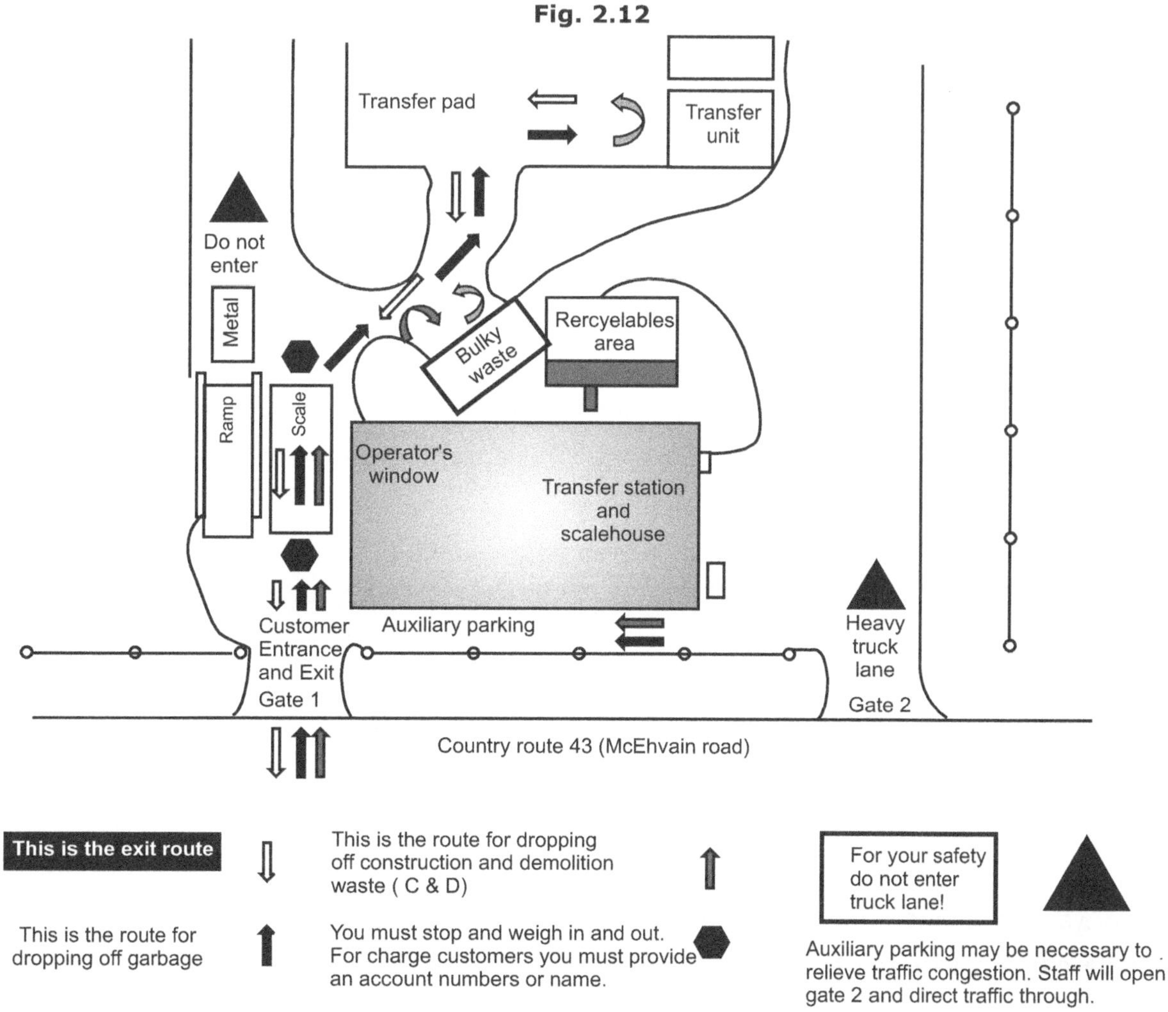

Fig. 2.13: Transfer Station Map

Location: (W-15)

- It should be centrally located in the town.
- It should not be far from disposal sites.
- It should be located in the area where traffic flow is smooth.
- It should be in area where land and property rates are cheap.
- It should not be in the thickly populated area or should not be in residential area at all as far as possible.

2.5.8.3 Necessity of Transfer Station (W-15, 16; S-15)

Transfer stations are necessary due to following reasons:

1. They prevent the scattering of MSW.
2. To have ease in proper storage and collection of MSW from different locations.
3. To prevent nuisance due to scattered solid waste to nearby area.
4. To reduce the haul distance.

- With the large volume of solid waste generated from urban markets, it is often not feasible or economical to transport waste directly from the collection point to the landfill, which is usually situated some distance away from the generation point.
- In these situations, it is wise to have Transfer Stations strategically located to accept waste from collection trucks. At these Transfer Stations, waste is consolidated, compacted and loaded into long-haul trailers for transport to landfills.
- This operation improves the utilization of collection equipment by minimizing transportation time and efficiently moving large volumes of waste to our disposal sites.

2.5.8.4 Factors Affecting Selection of Site for Transfer Station (S-15; W-18)

1. **Waste scattering/pollution:** It should be selected such that it should not create nuisance to nearby areas.
2. **Haul distance:** It should be constructed at suitable locations so as to minimize haul distances.
3. **Heritage place:** It should be away from heritage place.
4. **Traffic flow:** It should be located in area where traffic flow is smooth.
5. **Electricity:** Electricity should be available.
6. **Rate of land:** It should be located in area where rates of land and property are cheap.

2.6 ORGANIZATION PATTERN OF SOLID WASTE MANAGEMENT

(W-15, 18; S-16, 19)

- Organizational patterns are structures of relationship, usually in a professional organization, that help the organization to achieve its goals.
- It is necessary that, the local body takes adequate measures for institutional strengthening through induction of professionals, decentralization of administration, delegation of powers, human resources development and private sector and NGO participation for realizing the goals of solid waste management efficiently.

2.6.1 Decentralization of Administration

- In large cities, the SWM services can be performed effectively only if its administration is adequately decentralised.

- Decentralisation can be at least 3 tiered - One at the Ward level, second at the Zone level and third at the City level.
- The SWM functions would get focussed attention if all functions of the city administration are decentralised at Ward/Zone/Division levels and senior officers are placed in-charge of each Zone/Division functioning independently with adequate delegated powers.

2.6.2 Delegation of Powers

- Authority and responsibility should go hand in hand. For fixing accountability, there should be adequate delegation of fiscal and disciplinary powers to the officers and the supervisory staff responsible for managing solid waste and carrying out all day-to-day functions smoothly.
- The Head of the SWM department should also have the power to punish subordinates including supervisory staff. Adequate in-built checks may be introduced to ensure that the delegated powers are not misused.

2.6.3 Organizational Set-up

2.6.3.1 Towns Below 1,00,000 Population

- One qualified sanitation diploma holder as Sanitary officer (S.O.) if the population is more than 50000.
- One qualified Sanitary Inspector (S.I.) @ 50000 population.
- One qualified Sanitary Sub-Inspector (S.S.I) @ 25000 population.
- One Sanitary Supervisor (S.S.) @ 12,500 population.

2.6.3.2 Cities Between 1 and 2 Lacs Population

- Public Health/Environmental Engineer or Civil Engineer having training in environmental/public health engineering in the grade of Assistant Engineer to be in-charge of SWM department.
- Qualified Sanitation Diploma holder/Sanitation Officer @ 1 S.O. per 1 lac population or part thereof to look-after the collection, transportation, processing and disposal of waste or @ 1 per 2 Sanitary inspectors, whichever is less.
- Qualified Sanitation Diploma holder/Sanitary Inspector (S.I.) @ 1 S.I. per 50,000 population or part thereof or @ 1 per 80 sweepers, whichever is less.
- Qualified sanitation diploma holder/Sanitary Sub-inspector (S.S.I.) @ 1 S.S.I. per 25,000 population or part thereof or @ 1 per 40 sweepers, whichever is less.
- Sanitary Supervisors (a person who can read, write and report) @ 1 S. S. per 12,500 population or part thereof, or 1 per 20 sweepers, whichever is less.

2.6.3.3 Cities Having Population Between 2 and 5 Lacs

- Public Health/Environmental Engineer or Civil Engineer having training in environmental or public health engineering in the Grade of Assistant Executive Engineer to be in charge of SWM department.
- Public Health/Environmental Engineer in the grade of Assistant Engineer to look after the transportation, processing and disposal of waste.
- Sanitary Officers, @ one S.O. per 1 lac population for supervising the storage, street sweeping and primary collection of waste.
- Sanitary Inspectors, Sanitary Sub-inspectors, Sanitary Supervisors should be as per the yardstick indicated in 2.6.3.2 above.

2.6.3.4 Cities Having Population Between 5 and 20 Lacs (S-15)

- Public Health/Environmental Engineer or Civil Engineer having training in environmental or public health engineering of the level of Executive Engineer to be in-charge of the SWM department.
- Public Health/Environmental Engineers or Civil Engineer having training in environmental or public health engineering of the level of Assistant Executive Engineer per 5 lacs population.
- Public Health/Environmental Engineers or Civil Engineer having training in environmental or public health engineering of the level of Assistant Engineer per 2.5 lacs population.
- Sanitary Officers, Sanitary Inspectors, Sanitary Sub-inspectors and Sanitary Supervisors as per yardstick indicated in 2.6.3.2 above.

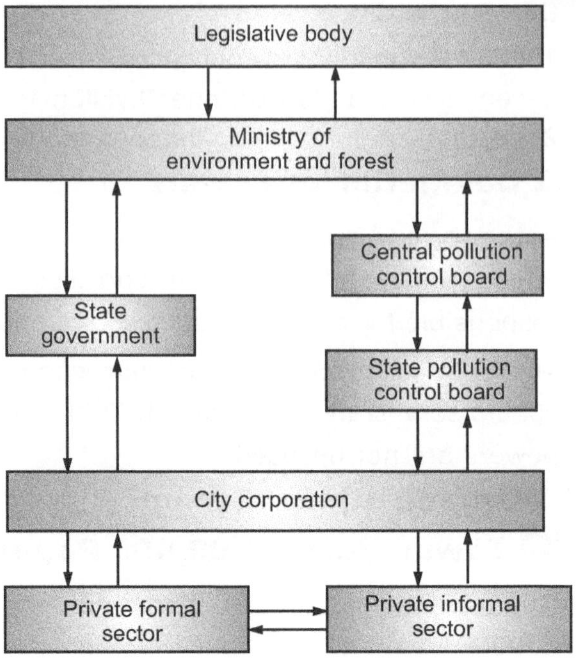

Fig. 2.14: Organisational Set-up of Solid Waste Management (SWM) Administration (W-15, 16)

2.6.3.5 Cities Having Population between 20 and 50 Lacs

- Public Health Engineer/Environmental Engineer or Civil Engineer having training in environmental or public health engineering of the level of Superintending Engineer to be the Head of SWM Department.
- Public Health/Environmental Engineers or Civil Engineer having training in environmental or public health engineering of the level of Executive Engineer @ 1 Executive Engineer per 20 lacs population or part thereof.
- Rest of the supervisors and staff as per the yardstick already indicated in 2.6.3.2 to 2.6.3.4 above.

2.6.3.6 Cities Having Population Above 50 Lacs

- Public Health Engineer/Environmental Engineer or Civil Engineer having training in environmental or public health engineering of the level of Chief Engineer to be in-charge of the SWM department.
- Superintending Engineer per 40 lac population or part thereof. Rest of the officers, supervisor's etc. as per yardsticks already indicated in 2.6.3.4 to 2.6.3.5 above.

2.7 IMPORTANT ORGANISATIONS TO CONTROL SOLID WASTE MANAGEMENT (S-19)

(i) Ministry of Environment and Forests: Ministry of Environment and Forests is responsible for all of the environmental policy at the national level, including the management of waste. The Ministry has an overview of all the activities of the MSWM sector and makes sure that it is performed well.

(ii) Central Pollution Control Board: Central Pollution Control Board keeps a check on all the activities that have potential to pollute the environment, which includes the monitoring of the municipal solid waste management in the country. It has divisions in each state that report to CPCB on the environmentally hazardous activities in the state, the actions taken towards them and the improvements made by the industries and public towards a cleaner environment.

(iii) State Pollution Control Board: State Pollution Control Board keeps a check on all the activities that have potential to pollute the environment, which includes the monitoring of the municipal solid waste management in the state. It reviews the Environmental Impact Assessment carried out by the agencies prior to the construction of a landfill site, installation of an incinerator or any other processing plant. It carries out public participation meeting to make the public aware of the proposed project and its benefits. Public participation is especially important so that once the project is started there should not be any agitation against the project.

(iv) City Corporation: City Corporation is responsible for the making of the solid waste management policy, setting up the targets and objectives. They are responsible for managing the solid waste in the city and are answerable to the State Board and Central Pollution Control Board. They also have the authority to privatize the solid waste management sector.

(v) Private formal sector: Their functions include the collection of waste, transfer of waste to trucks, transport of waste to the specified dump yard. They have the responsibility of disposing off the waste by alternative layering of waste and soil, spraying it with EM solution and water.

(vi) Private informal sector: The informal sector in the city is very large and plays a very vital role in the MSWM. It comprises of the rag pickers who retrieve recyclable waste from the community bins and landfills, and sell it to either bigger dealers or to recycling factories.

Health Aspects Involved in Handling and Processing of Solid Waste: **(S-19)**

- There is potential risk to environment and health from improper handling of solid wastes. Direct health risks concern mainly the workers in this field, who need to be protected, as far as possible, from contact with waste.
- Traffic accidents can result from toxic spilled wastes.
- Air pollution can be caused from the inefficient burning of wastes, either in open air, or in plants that lack effective treatment facilities from the gaseous effluents.
- Uncontrolled hazardous wastes from industries mixing up with municipal wastes create potential risk to human health.
- The most obvious environmental damage caused by municipal solid wastes is aesthetic, the ugliness of street litter and degradation of urban environment and beauty of city.
- There is specific danger of concentration of heavy metals in the food chain, a problem that illustrates the relationship between municipal solid wastes and liquid industrial effluents containing heavy metals discharged to a drainage/ sewerage system and/ or open dumping sites of municipal solid wastes and the wastes discharged thereby maintains a vicious cycle.
- Municipal Solid Wastes Management Systems involves various activities like storage, collection, transportation, disposal etc. These activities even if properly controlled and with proper precautionary measures adopted, may have adverse impact on land, water and air environment, human and environmental health aesthetics and quality of life.
- The main risk to health is indirect and arises from the breeding of disease vectors, primarily files and rats.
- Biodegradation of organic matter may pose health problems.
- Mixing of Bio medical waste, Industrial waste in Municipal waste causes health hazards.

Important Points

- Storage of waste is the first step of SWM.
- Collection of waste is the second step of SWM.
- Solid waste are transported in bulk to the processing or disposal sites from waste storage depot.
- Use of non-corrosive container with lid is advised for storage of waste.
- Litter bins may be provided on important streets, markets, public places etc.
- Door step collection is one of the method of collection of waste.
- Litter bins, road side litter bin, broom, shovels, hand cart, mechanical road sweeper, community bins are the tools and equipments required for SWM.
- Animal carts, auto vehicles, tractors, trucks, dumpers, compactor vehicles are used for transportation of solid waste.
- A transfer station is a building or processing site for the temporary deposition of waste.
- In large cities, 3 tiered administration levels are set up for proper management of solid waste.
- For the effective management of solid waste, there is organizational and administrative set-up as per the population of the town or city.

Practice Questions

1. Define storage, collection and transportation of solid waste.
2. Describe the present scenario with regard to collection of solid waste.
3. How is the storage of the following done:
   (a) Household Waste
   (b) Waste from Shops/ Offices/ Institutions/ Workshops
   (c) Waste from Hotels and Restaurants
   (d) Vegetable/Fruit Markets Waste
   (e) Meat and Fish Markets
   (f) Street Food Vendors Waste
   (g) Marriage Halls/Kalyan Mandaps/Community Halls etc. Waste
   (h) Hospitals / Nursing Homes / Pathological Laboratories / Health Care Centres / Establishments etc.
   (i) Construction and Demolition Wastes
   (j) Garden Waste
   (k) Recyclable Waste
4. What are the measures to be taken by the local bodies towards segregation of recyclable waste?
5. How are the provisions of litter bins on streets, public places etc. done?
6. How are the provisions of special containers for storage of domestic hazardous and toxic waste done?
7. Describe the present scenario with regard to collection of municipal waste.
8. What are the measures to be taken to improve the collection services of municipal waste?
9. What are the arrangements to be made for the efficient collection of waste?

10. Describe the following methods of collection of waste:
    (a) Collection by Motorised Vehicles
    (b) Door Step Collection through Containerized Handcarts
    (c) Collection of Waste from Societies/Complexes
    (d) Collection of Waste from Slums
    (e) Collection-at-the Doorstep in Posh Residential Areas
    (f) Collection of dairy and cattle-shed waste
    (g) Collection of domestic hazardous and toxic waste
    (h) Collection of Duly segregated recyclable/Non-bio-degradable waste from Households
    (i) Collection of waste from shops and establishments
    (j) Collection of bio-medical waste
    (k) Collection of hotel and restaurant waste
    (l) Collection of construction and demolition waste
    (m) Collection of waste from marriage halls, kalyan mandaps, community halls etc.
    (n) Collection of waste from gardens
    (o) Collection of waste from vegetable, fruits, meat and fish markets
11. Describe the following tools and equipments:
    (a) Litter Bin
    (b) Broom
    (c) Shovels
    (d) Handcarts
    (e) Mechanical Road Sweeper
    (f) Community Bins (movable and stationary)
11. Write down in detail the present scenario/problems in transportation of waste.
12. What are the measures to be taken to improve the transportation system for SWM?
13. Write a short note on transportation vehicles with their capacities and working.
14. Explain the meaning of transfer station, its necessity and location.
15. What is organization pattern of solid waste management?
16. Explain:
    (a) Decentralization of administration
    (b) Delegation of powers
    (c) Organizational set-up

MSBTE Questions and Answers (As Per 'G' Scheme)

Summer 2016

1. State various methods of storage of municipal solid waste. **(4M)**

Ans. Refer Section 2.1.1

2. List the various transportation equipment of municipal waste. **(4M)**

Ans. Refer Section 2.5.

3. What is transfer station and state its necessity. **(4M)**

Ans. Refer Sections 2.5.8 and 2.5.8.3.

4. Enlist collection methods of municipal solid waste. Explain any one. **(4M)**

Ans. Refer Section 2.2.5.

Winter 2016

1. Enlist any four transport vehicles with their capacities for transporting municipal waste. **(4M)**

Ans. Refer Section 2.5.

2. Enlist various methods of collection of municipal solid waste and explain any one of them. **(4M)**

Ans. Refer Section 2.2.5.

3. Draw the organization pattern of solid waste management. **(4M)**

Ans. Refer Section 2.6.

4. Explain in brief transfer station. **(4M)**

Ans. Refer Section 2.5.8.

5. State the methods of storage of solid municipal waste. **(4M)**

Ans. Refer Section 2.1.

Summer 2017

1. Draw the organisation pattern of Solid Waste Management. **(4M)**

Ans. Refer Section 2.6.

2. List the various eight types of collection and transportation equipments of Solid Waste. **(4M)**

Ans. Refer Section 2.3.

3. State various methods of storage of Municipal Solid Waste. **(4M)**

Ans. Refer Section 2.1.

4. Explain any two methods of collection of municipal waste. **(4M)**

Ans. Refer Section 2.2.

5. What are the different types of health problems during time of segregation, reuse, recovery and recycling of solid waste possible to human being ? **(4M)**

Ans. Refer Section 2.2.6.

6. What is meaning of transfer station ? And explain its necessity and suitable location. **(4M)**

Ans. Refer Section 2.5.8.

Winter 2017

1. List methods of collection of municipal solid waste. **(2M)**

Ans. Refer Section 2.2.2.

2. Write any two precautions to be taken during handling and processing of solid waste. **(2M)**

Ans. Refer Section 2.2.4.9.

3. Define transfer station. State necessity of transfer station. State the factors affecting selection of site for transfer station. **(4M)**

Ans. Refer Section 2.5.8.

4. State the methods of storage of (MSW) municipal solid waste. State current practice of solid waste management in India in brief. **(8M)**

Ans. Refer Sections 2.1.

5. Draw the organisation pattern of solid waste management. **(4M)**

Ans. Refer Fig. 2.6.

Summer 2018

1. Explain in brief about organisation pattern of solid waste management according to the population of city. **(4M)**

Ans. Refer Section 2.6.

2. Explain in brief the present scenario with regard to collection of municipal solid waste. **(4M)**

Ans. Refer Section 2.2.1.

3. State the method of storage for following types of waste. **(4M)**

 (i) Vegetable/fruit market waste, (ii) Street food venders waste,

 (iii) Household waste, (iv) Garden waste.

Ans. Refer Section 2.1.

4. State the methods of provisions of Litter bin on streets and public places. **(4M)**

Ans. Refer Section 2.3.1.

5. Explain the health aspects during handling and processing of solid waste. **(4M)**

Ans. Refer Section 2.2.5.

6. Describe about mechanical road sweepers used in SWM practices with sketch. **(4M)**

Ans. Refer Section 2.3.5.

7. Define transfer station. State about its necessity and criteria of location of transfer station. **(4M)**

Ans. Refer Section 2.5.8.

Winter 2018

1. Explain the methods of collection of municipal solid waste. **(4M)**

Ans. Refer Section 2.2.2.

2. Explain transportation system for municipal solid waste. **(4M)**

Ans. Refer Section 2.4.

3. Draw the organization pattern of solid waste management. **(4M)**

Ans. Refer Fig. 2.6.

4. State requirements of location of transfer station. **(4M)**

Ans. Refer Section 2.5.8.2.

5. Draw neat labeled sketch of Litter Bin. **(4M)**

Ans. Refer Section 2.3.1.

6. State the parameters to be taken into account while making storage facility for solid waste. **(4M)**

Ans. Refer Section 2.1.13.

7. Explain health problems during segregation and recycling of solid waste. **(4M)**

Ans. Refer Section 2.2.5.

Summer 2019

1. What are the measures to be taken towards the segregation of recyclable waste? **(4M)**

Ans. Refer Section 2.2.5.9.

2. State the measures to be taken to improve the collection services of municipal waste. **(4M)**

Ans. Refer Section 2.2.3.

3. Describe any two methods of collection of waste from the following: **(8M)**
   (i) Collection of bio-medical waste.
   (ii) Collection of waste from vegetable, fruits, meat and fish markets.
   (iii) Collection of construction and demolition waste.

Ans. Refer Sections 2.2.5.11, 2.2.5.16 and 2.2.5.13.

4. Explain the meaning of transfer station, its necessity and location. **(4M)**

Ans. Refer Section 2.5.8.

5. Describe the organization pattern of solid waste management. **(4M)**

Ans. Refer Section 2.6.

6. What is waste segregation? Explain its importance. **(4M)**

Ans. Refer Section 2.2.5.9.

7. State how enforcement can help in stopping littering in the street. **(4M)**

Ans. Refer Section 2.3.1.

8. Describe the health aspect during handling and processing of waste. **(4M)**

Ans. Refer Chapter 2, Section 2.2.5.

***

Chapter **3...**

Disposal of Municipal Solid Waste

Weightage of Marks = 16, Teaching Hours = 14

Syllabus

3.1 Concept of Composting of Waste, Principles of Composting Process. Factors Affecting the Composting Process

3.2 Methods of Composting

- (A) Manual Composting – Bangalore Method, Indore Method
- (B) Mechanical Composting – Dano Process
- (C) Vermicomposting

3.3 Land Filling Technique, Factors to be Considered for Site Selection

- Land Filling Methods - Area Method, Trench Method and Ramp Method.
- Leachate and Its Control, Biogas from Landfill
- Advantages and Disadvantages of Landfill Method
- Recycling of Municipal Solid Waste

3.4 Incineration of Waste

- Introduction of Incineration Process
- Types of Incinerators - Flash, Multiple Chamber Incinerator
- Products of Incineration Process with their Use
- Pyrolysis of Waste - Definitions, Methods

3.5 Products of Incineration Process

- Advantages and Disadvantages of Incineration Process

About this Chapter

At the end of this chapter, students will be able to:

- Explain the principles of preparing the compost for the given site.
- List the relevant factors affecting the given composting process with justification.
- Describe the different steps of executing the relevant method of composting for the given site.
- Suggest the design criteria adopted in execution of vermicomposting for the given area with justification.
- Explain the relevant operating method of sanitary land filling for the given site condition and given type of waste.
- Select the site suitable for the land filling purpose in the given locality with justification.
- State the relevant parameters to select a site for land filling with justification.
- Propose the relevant method to control the liquid Leachate generated in the given land filling site.

3.0 OVERVIEW

- Disposal of solid wastes on land is by far the most common method in most of the countries and probably accounts for more than 90% of the world's municipal solid waste.
- Incineration accounts for most of the remainder, whereas composting of solid wastes accounts for only an insignificant amount.
- Selecting a disposal method depends almost entirely on costs, which in turn are likely to reflect local circumstances.

3.1 COMPOSTING OF WASTE

- The organic content of Municipal Solid Waste (MSW) tends to decompose leading to various smell and odour problems. It also leads to pollution of the environment. To ensure a safe disposal of the MSW, it is desirable to reduce its pollution potential, and several processing methods are proposed for this purpose.
- Composting process is quite commonly used and results in production of a stable product - compost, which depending upon its quality can be used as a low grade manure and soil conditioner. The process results in conservation of natural resources and is an important processing method, especially in agricultural and horticultural areas.
- In the case of individual households, small establishments and colonies, vermi-composting which involves the stabilisation of organic solid waste through earthworm consumption for conversion of the organic material to worm casting is being increasingly preferred.

3.1.1 Principles of Composting (S-15, 19; W-15, 16)

- Composting is the biological decomposition of organic waste such as food or plant material by bacteria, fungi, worms and other organisms under controlled aerobic (occurring in the presence of oxygen) conditions. The end result of composting is an accumulation of partially decayed organic matter called humus.
- Decomposition and stabilization of organic waste matter is a natural phenomenon. Composting is an organised method of producing compost manure by adopting this natural phenomenon.
- Compost is particularly useful as organic manure which contains plant nutrients (Nitrogen, Phosphorous and Potassium) as well as micro-nutrients which can be utilized for the growth of plants. When used in conjunction with chemical fertilisers optimum results are obtained.
- Composting can be carried out in two ways i.e., aerobically and anaerobically.
  - **(i) Acrobic composting :** During aerobic composting aerobic micro-organisms oxidise organic compounds to Carbondioxide, Nitrite and Nitrate. Carbon from organic compounds is used as a source of energy while nitrogen is recycled. Due to exothermic reaction, temperature of the mass rises.
  - **(ii) Anaerobic composting:** During anaerobic process, the anaerobic micro-organisms, while metabolising the nutrients, break down the organic compounds through a process of reduction. A very small amount of energy is released during the process and the temperature of composting mass does not rise much. The gases evolved are mainly Methane and Carbondioxide.
- The end result of composting is an accumulation of partially decayed organic matter called humus.
- An anaerobic process is a reduction process and the final product is subjected to some minor oxidation when applied to land.

3.1.2 Factors Affecting the Composting Process (S-15, 16; W-16, 18)

Factors Affecting the Composting Process:

1. Organisms
2. Moisture
3. Temperature
4. Aeration
5. Addition of sewage and sewage sludge
6. Use of cultures
7. C/N ratio
8. Particle size
9. pH
10. Blending and seeding
11. Air circulation.

3.1.2.1 Organisms

- Composting of organic waste depends upon the type of bacteria involved in composting process.
- Aerobic composting is a dynamic system wherein bacteria, actionoycetes, fungi and other biological forms are actively involved. The relative importance of one species over another depends upon the constantly changing food supply, temperature and substrate conditions.
- Facultative and obligate forms of bacteria, actinomycetes and fungi are most active in this process. In the initial stages, mesophilic forms predominate and thermophilic bacteria and fungi soon take over except in the final stage of composting.
- Except when the temperature drops, actinomycetes and fungi are confined to 5 to 15 cm outer surface layer. If the turning is not carried out frequently, the actinomycetes and fungi in these layers register increased growth imparting it typical greyish white colour. Thermophilic actinomycetes and fungi are known to grow well in the range of 45° to 60°C.
- Different organisms are known to play predominant role in breaking down different constituents of municipal solid waste. Thermophilic bacteria are mainly responsible for the breakdown of proteins and other readily biodegradable organic matter.
- Fungi and actinomycetes play an important role in the decomposition of cellulose and lignin. The actinomycetes common in compost are Streptomyces species and Micromonospora species, the latter being more prevalent. The common fungi in compost are Thermonomyces species, Penicillium dupontii and Asperigallus fumigatus.
- Majority of these organisms responsible for composting are already present in municipal solid waste. Not much information is available regarding the organisms active in anaerobic composting, though many of the organisms responsible for anaerobic decomposition of sewage sludge will be active here also and differences are expected due to the concentration of nutrients present and the temperature conditions.

3.1.2.2 Moisture

- The moisture tends to occupy the free air space between the particles. Hence, when the moisture content is very high, anaerobic conditions set in.
- However, the composting mass should have a certain minimum moisture content in it for the organisms to survive. The optimum moisture content is known to be between 50 to 60%.
- Higher moisture content may be required while composting straw and strong fibrous material which soften the fibre and fills the large pore spaces. Higher moisture content can also be used in mechanically aerated digesters.
- In anaerobic composting, the moisture content used will depend upon the method of handling and whether it is carried out in the open or in closed container.

3.1.2.3 Temperature

- Activities of organism depends upon the temperature, so for efficient composting process optimum temperature need to be maintained.
- The aerobic decomposition of a gram mole of glucose releases 484 to 674 kilo calories (kcal) energy under controlled conditions, while only 26 kcal are released when it is decomposed anaerobically.
- Municipal solid waste is known to have good insulation properties and hence the released heat results in increase in temperature of the decomposing mass. As some of the heat loss occurs from the exposed surface, the actual rise in temperature will be slightly less. When the decomposing mass is disturbed, as during turning of windrows, the resultant heat loss results in drop in temperatures.
- Under properly controlled conditions temperatures are known to rise beyond 70°C in aerobic composting. During anaerobic composting as the released heat is quite small and as part of it is lost from the surface, only a marginal rise in temperature occurs.

- This increased temperature results in increased rate of biological activity and hence results in faster stabilisation of the material. However, if the temperature rise is very high, due to inactivation of the organisms and enzymes, the rate of activity may decrease.
- The studies carried out have shown that the activity of cellulose enzyme reduces above 70°C and the optimum temperature range for nitrification is 30° to 50°C beyond which nitrogen loss is known to occur. The temperature range of 50° to 60°C is thus optimum for nitrification and cellulose degradation.
- The high temperature also helps in destruction of some common pathogens and parasites. The studies carried out at NEERI have shown that the destruction of these organisms is not ensured under anaerobic conditions.
- It has been proved that the high temperature and long retention during aerobic composting along with the antibiotic effect results in destruction of parasites and pathogens.
- Thus, if the process is so controlled that the temperature is kept between 50° to 60°C for 5 to 7 days, destruction of pathogens and parasites can be ensured.

3.1.2.4 Aeration

- It is necessary to ensure that oxygen is supplied throughout the mass and aerobic activity is maintained. During the decomposition, the oxygen gets depleted and has to be continuously replenished.
- This can be achieved either by turning of windrows or by supplying compressed air. During the turning, it is necessary to bring inner mass to the outer surface and to transfer the outer waste to the inner portion.
- In case of artificial air supply, the quantity of air supply is normally maintained at 1-2 cu.m./day/kg of volatile solids. Artificial air supply requires enclosing decomposing mass in containers which is quite costly.
- Hence in Indian conditions, the decomposition is commonly carried out in open windrows. Studies at NEERI have shown that the optimum turning interval which will reduce the cost and simultaneously maintain aerobic conditions is 5 days.

3.1.2.5 Addition of Sewage and Sewage Sludge

- To maintain optimum C/N ratio, addition of sewage is done as it also bring down the cost of sewage sludge treatment.
- Addition of sewage sludge increases smell and odour problems. It will also increase handling and transportation cost. Even if sewage is used as a source of moisture, bulk of sewage will still have to be treated.
- The sewage often contains waste waters from industries which contain hazardous constituents which will pose problems in the composting process and compost quality. In view of the above, addition of sewage and sludge is not desirable in India.

3.1.2.6 Use of Cultures

- Addition of similar and extraneous organisms in the form of inoculum is unnecessary in Indian conditions.
- However, such inoculum will be required during composting of industrial and agricultural solid waste which do not have the large mix of indigenous bacterial population.

3.1.2.7 Carbon/Nitrogen (C/N) Ratio (W-15)

- The optimum C/N ratio for composting is 25-30 depending upon other environmental conditions. MSW in developed countries has a C/N ratio of nearly 80. To bring it down to the optimum value and to reduce the cost of sewage sludge treatment, it is mixed with sewage sludge (C/N ratio = 5 to 8).

- MSW in India, on the other hand has an initial C/N ratio of around 30 which does not need blending. If such a mixing is done, C/N value may reduce below 20, when a loss of nitrogen in the form of ammonia occurs.
- The organisms involved in stabilisation of organic matter utilise about 30 parts of carbon for each part of nitrogen and hence an initial C/N ratio of 30 is most favourable for composting.
- Research workers have reported the optimum value to range between 26-31 depending upon other environmental conditions.
- The C/N ratio considers the available carbon as well as the available nitrogen while the available carbon and nitrogen in the MSW may vary from sample to sample.
- Whenever the C/N ratio is less than the optimum, carbon source such as straw, sawdust, paper are added while if the ratio is too high, the sewage sludge, slaughter house waste, blood etc. are added as a source of nitrogen.

3.1.2.8 Bending and Seeding

- It is required to maintain the moisture content and to accelerate biological decomposition of waste and hence affect the rate of composting.
- Proper timing should be followed for blending and seeding.

3.1.3 Favourable Conditions for Composting (S-17)

- Certain temperatures promote rapid composting and destroy pathogens and weed seeds.
- Microorganisms living in a compost pile need an adequate amount of moisture to survive.
- Providing proper ventilation by designing the composting toilet with a vent pipe.
- The proper balance of nutrients is vital to the composting process. A good carbon/nitrogen (C/N) ratio for a compost toilet is 20/1 to 35/1.

3.2 METHODS OF COMPOSTING

3.2.1 Manual Composting

- These are the composting methods where no mechanical means are used; only human beings are used as labour/energy giving entities. These are mainly of two types: (i) Bangalore Method, (ii) Indore Method.

3.2.1.1 Bangalore Method (Anaerobic Method) (W-15, 18; S-18, 19)

- This is an anaerobic method conventionally carried out in pits.
- Formerly the waste was anaerobically stabilised in pits where alternate layers of MSW and night soil were laid.
- The pit is completely filled and a final soil layer is laid to prevent fly breeding, entry of rain water into the pit and for conservation of the released energy.
- The material is allowed to decompose for 4 to 6 months after which the stabilised material is taken out and used as compost.
- The Bangalore method requires longer time for stablisation of the material and hence needs larger land space, which is in short supply in urban areas. The gases generated in this anaerobic process also pose smell and odour problems.

Procedure:

- Bangalore method is commonly used anaerobic method used for biological conversion of organic component of municipal solid waste.
- In this method, underground earthen trench is excavated and alternate layer of waste and soil is filled in trench or pit to control odour. Final layer of soil is provided at the top.
- The soil cover not only prevent odour but also prevents breeding of files.
- Within 2-3 days of burial, intensive biological action starts taking place and organic matter begins to be destroyed. After 4-5 months complete stabilization of waste takes place.
- During biological action, heat is evolved which rises temperature of decomposing mass.

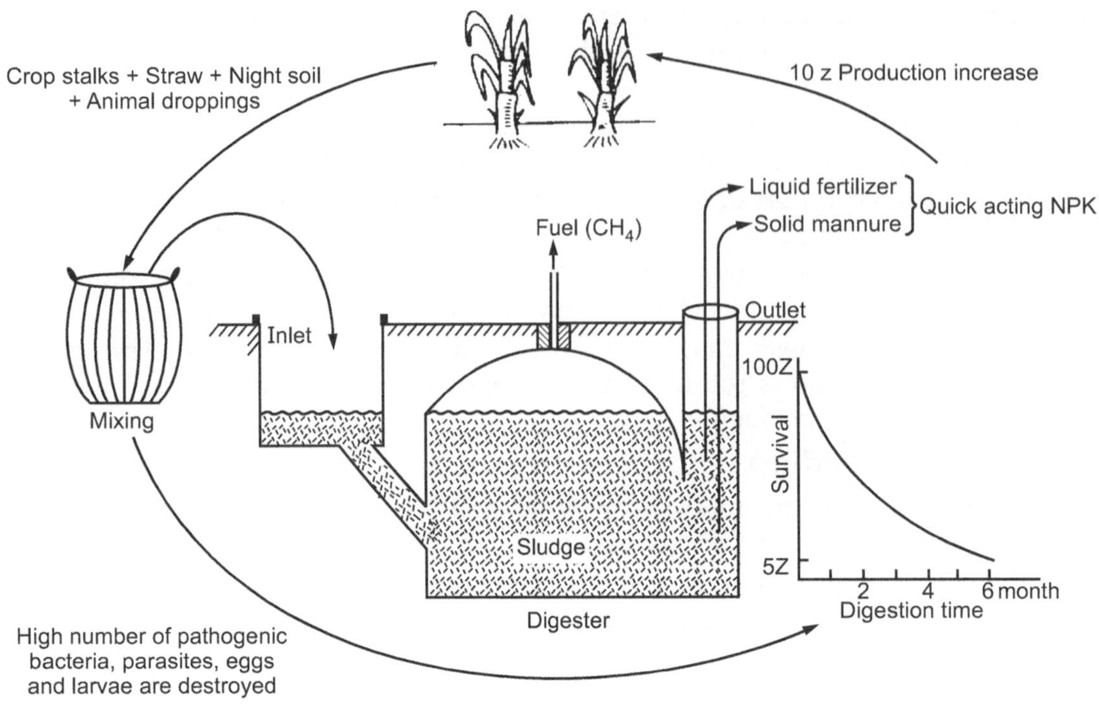

$$C_6H_{10}O_5 + H_2O \xrightarrow{\text{Anaerobic fermentation}} 3CH_4 + 3CO_2$$

Fig. 3.1: Bangalore Method of Compost

3.2.1.2 Indore Method (Aerobic Method) (W-15; S-16)

- This method of composting in pits involves filling of alternate layers of similar thickness as in Bangalore method. However, to ensure aerobic condition, the material is turned at specific intervals for which a 60 cm strip on the longitudinal side of the pit is kept vacant.
- For starting the turning operation, the first turn is manually given using long handled rakes after 4 to 7 days of filling. The second turn is given after 5 to 10 more days. Further turning is normally not required and the compost is ready in 2 to 4 weeks.
- In the urban areas, due to extensive provision of water carriage system of sanitation, night soil is not available. Composting of MSW alone is hence often carried out. Aerobic composting of MSW is commonly carried out in windrows.
- The Indore method stabilises the material in shorter time and needs lesser land space. As no odourous gases are generated in this process, it is environment friendly and hence commonly preferred.

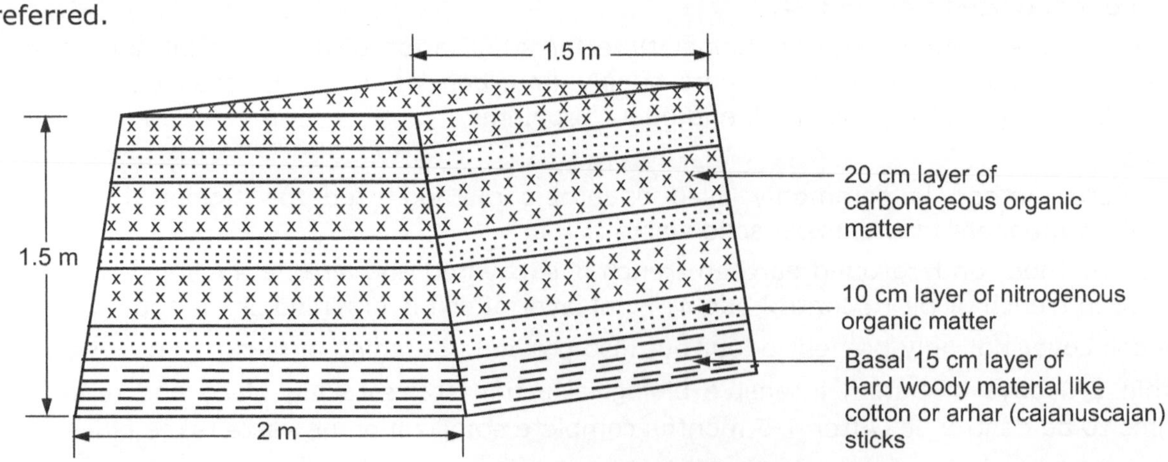

Fig. 3.2: Indore Method

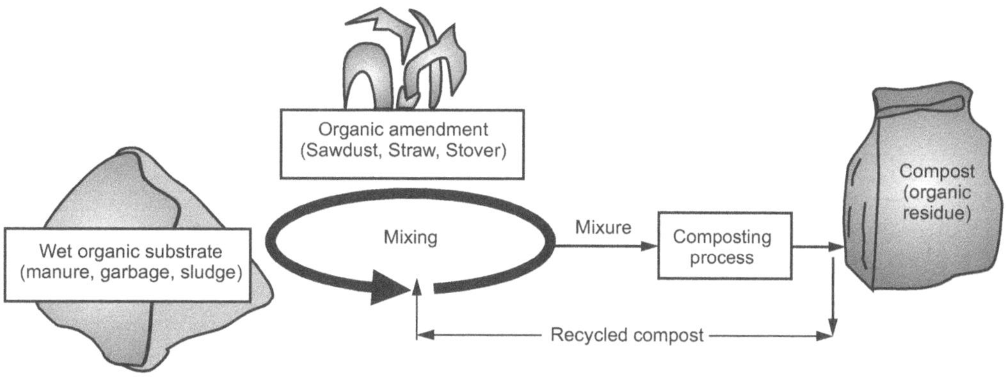

Fig. 3.3: Composting Process Materials Flow Diagram

Procedure:

- In Indore method layers of vegetable and night soil are alternatively piled into trench, the depth of pile is 1.5 - 2 m and width is about 3 - 8 m above the ground which form a mound called windrow.
- Normally, windrows are conical in shape and about 50 m in length.
- The aeration is achieved by periodically turning the piles. Manual turning is adopted for small plants and mechanical turning is adopted for larger plants.
- Refuse should be turn once or twice per week which introduce oxygen and helps to control temperature.
- Turning is continued for about 4 - 5 weeks during which biodegradable organic materials are consumed. The solid waste is allowed to keep for 2 - 8 weeks with turning. The composting in windrow may take 21 - 28 days for stabilization.
- The composted waste is removed from windrow and allowed to mature in maturing yards for 1 - 3 months, after which the compost becomes ready for being taken out for use.

3.2.1.3 Comparison between Bangalore Method and Indore Method (S-15, 19)

Bangalore method	Indore method
1. The Bangalore method functions aerobically for several days and then becomes anaerobic, because no turning occurs.	1. The Indore composting system relies on aerobic activity although portions of the pile or pit will likely become anaerobic between turnings.
2. Fly breeding and odour problems are often associated with this method.	2. This method has better fly control.
3. Slow method.	3. More rapid.
4. Non-uniform decomposition.	4. Uniform decomposition.
5. After 4 to 5 months the composting process is complete.	5. During the period of active composting, the contents are turned from 1 - 5 times over a period of 16 months.
6. More labour intensive method.	6. Less labour intensive method.
7. Moisture control problems.	7. Less moisture control problems.

3.2.2 Mechanical Composting - Dano Process (W-15)

- The process of stabilisation is expiated by mechanical device of turning the compost. It is recommended to use refuse of 1.5 cm particle size in this method. The moisture content and aeration of refuse are continually adjusted. Care is taken to see that quantity of air should not exceed 2 m^3/kg of volatile solid per day.
- It requires small area compared to trenching and open window composting.
- The stabilisation of waste takes 3 - 6 days.
- The operations involved are:
  - Reception of refuse
  - Segregation
  - Shredding
  - Stabilisation
  - Marketing.

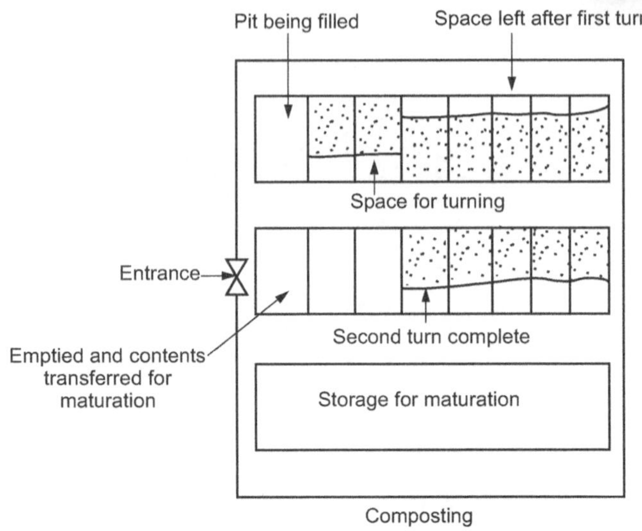

Fig. 3.4: Mechanical Composting (W-15)

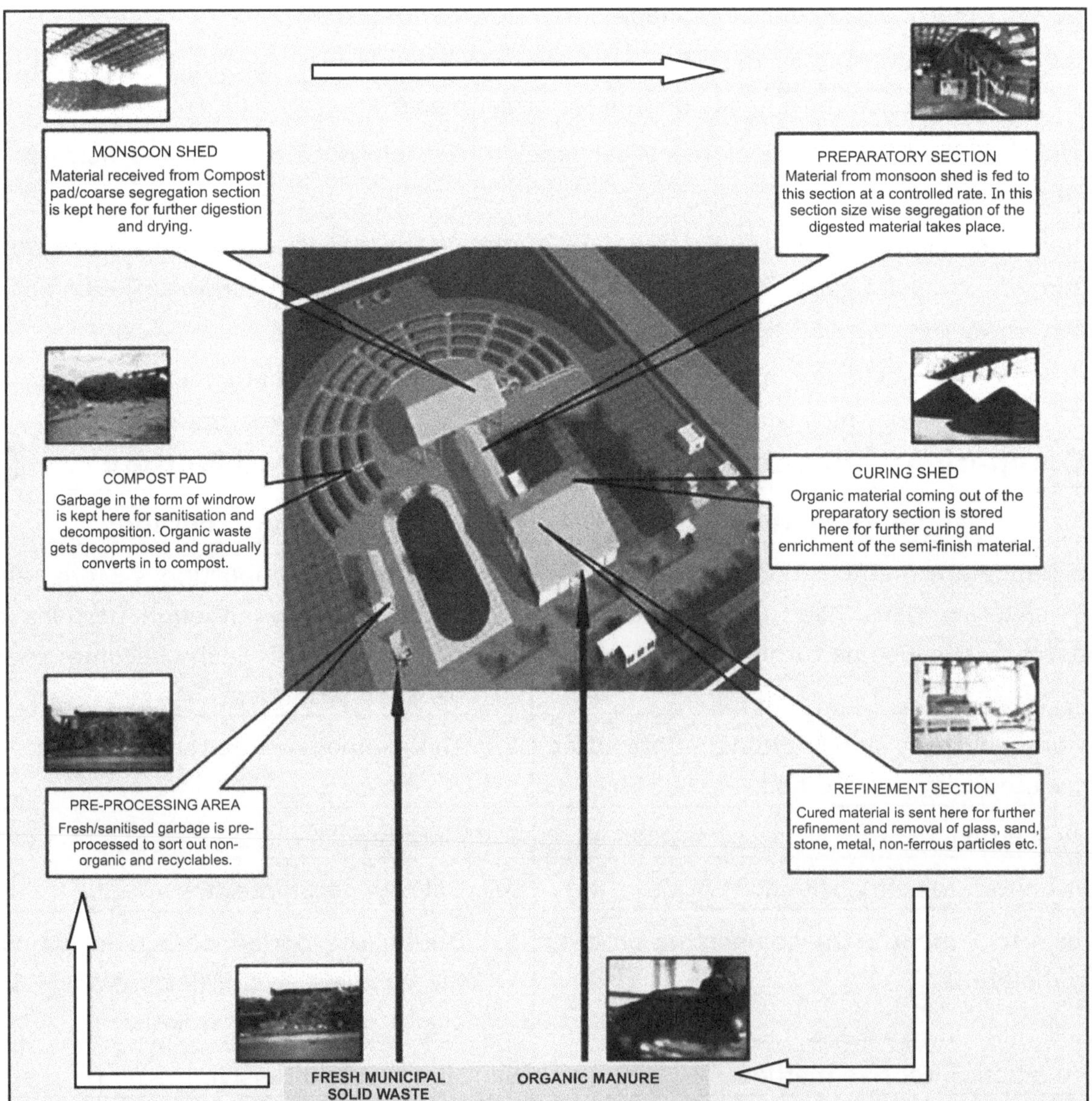

Fig. 3.5

- Most of the Municipal Solid Waste (MSW) received at compost plant is mixture of decomposable material with plastic, metal, glass, sand and other non-biodegradable material. MSW is first received in pre-processing area of the plant. In this phase, large size objects such as tyres, cans, glass articles, metals are hand picked manually.
- This pre-processed MSW is then shifted to composting pad for decomposition by making windrows. Composting culture is sprayed on this waste and each windrow is turned frequently to maintain aerobic conditions for composting with maximum speed. The concrete platform is provided with peripheral drain to collect the leachate and rain water overflow.

Fig. 3.6: Compost Pad

- Composted MSW is then shifted to coarse segregation area having mechanical arrangements to segregate the rejects like plastic, cloth, metals, glass of the size above 14 mm using rotary screen (trommel) of 35 mm and 15 mm sizes. Feeder, hopper and conveyer belts are provided with trommel for taking away rejects and semi-finished compost separately.

(a) Rotary Screen (Trommel) **(b) Series of Rotary Screen** **(c) Finished Compost**

Fig. 3.7

- Above semi-finished compost is then further shifted to curing section for further digestion. After curing, compost is shifted to refinement section which consists of vibratory screen or trommel of 4-6 mm size. Rejects like gravels, sand, stones, glass, metal and other inert materials above 4 mm are separated in this section. The fine compost manure is then collected and packed for selling.
- Trommel and conveyers provided in coarse segregation and refinement section operates on hydraulic system and electric motors with centralized control panel. These are designed to meet requirements of MSW management plants.
- Coarse segregation area, curing section and refinement section are under RCC shade with AC sheet roofing. This shade is properly ventilated and necessary sidecuts are provided to facilitate disposal of reject material.

Fig. 3.8: Segregated Waste

3.2.3 Vermicomposting (W-15, 16, 18; S-16, 18)

- Vermicompost is the product or process of composting using various worms, usually red wigglers, white worms and other earthworms to create a heterogeneous mixture of decomposing vegetable or food waste, bedding materials and vermicast. Vermicast, also called worm castings, worm humus or worm manure, is the end-product of the breakdown of organic matter by an earthworm. These castings have been shown to contain reduced levels of contaminants and a higher saturation of nutrients than do organic materials before vermicomposting.
- Vermicomposting involves the stabilisation of organic solid waste through earthworm consumption which converts the material into worm castings.
- Vermicomposting is the result of combined activity of microorganisms and earthworms. Microbial decomposition of biodegradable organic matter occurs through extracellular enzymatic activities (primary decomposition) whereas decomposition in earthworm occurs in alimentary tract by microorganisms inhabiting the gut (secondary decomposition). Microbes such as fungi, actinomycetes, protozoa etc. are reported to inhabit the gut of earthworms.
- Ingested feed substrates are subjected to grinding in the anterior part of the worms gut (gizzard) resulting in particle size reduction.
- Vermitechnology, a tripartite system which involves biomass, microbes and earthworms is influenced by the abiotic factors such as temperature, moisture, aeration etc. Microbial ecology changes according to change of abiotic factors in the biomass but decomposition never ceases. Conditions unfavourable to aerobic decomposition result in mortality of earthworms and subsequently no vermicomposting occurs.
- Hence, pre-processing of the waste as well as providing favourable environmental condition is necessary for vermicomposting.
- Vermicomposting is the breaking down of organic material through the use of worms, bacteria and fungi. In nature, organic matter is decomposed through these organisms. By managing vermicomposting we can essentially speed up mother nature's process of breaking down organic matter.

- The end product of vermicomposting is a substance called vermicompost or "worm castings".
- This is a nutrient rich organic substance that can be added to soil to increase it's organic matter content and available nutrients.
- A common way to use this process is to build "worm bins"; basically boxes that contain worms. Organic matter is added to the bin for the worms to eat! An advantage of "worm bins" over traditional composting is that, if they are inside during the winter, they allow this process to work all year.
- Using earthworms to convert organic wastes is an ecologically safe method that leads to an environmentally safe product. Two types of earthworms are recommended for this, they are: the Red Wiggler or manure worm (*Eisensia foetida*) and the Red Worm, another manure worm (*Lumbricus Rebellus*).

Fig. 3.9: Vermicompost

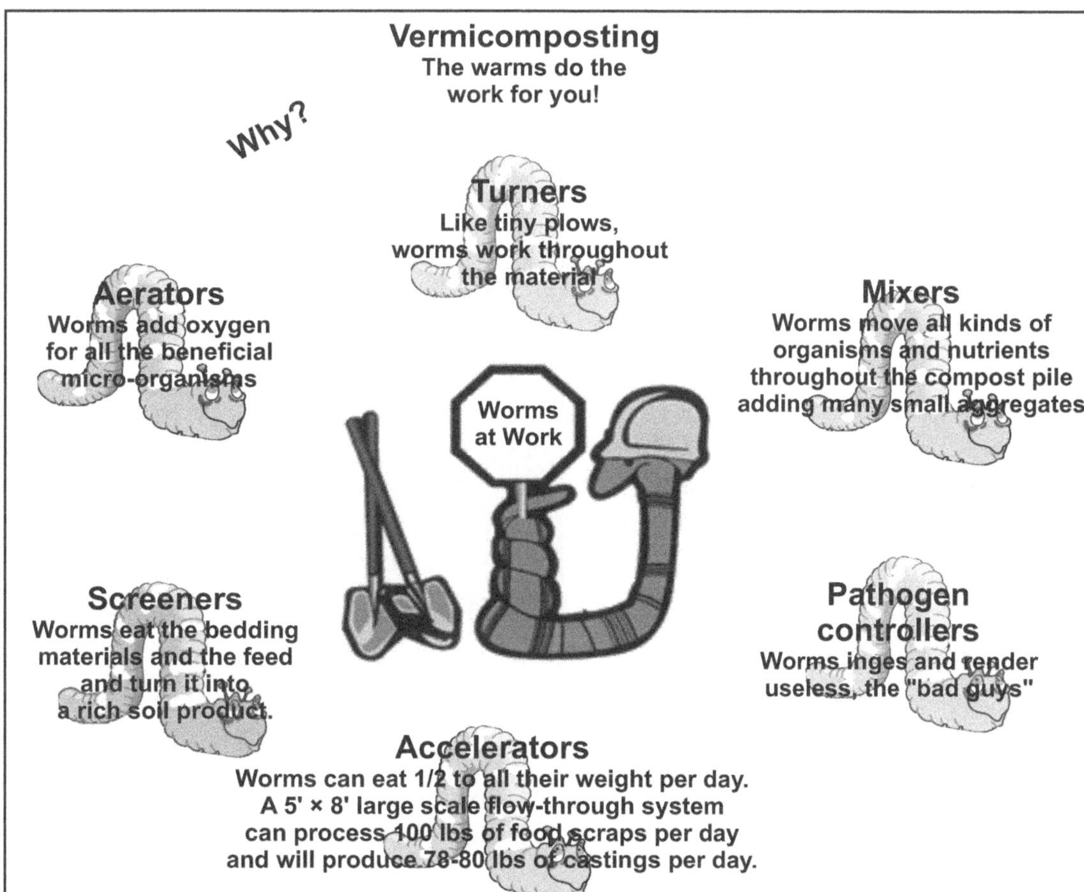

Fig. 3.10

Steps in Process of Vermicomposting: **(W-16)**

(i) Take a small wooden box or dig a small pit.

(ii) Spread a net on box and spread 1 or 2 cm thick layer of sand.

(iii) Put some cow dung and kitchen wastes such as remaining part of cut vevetables, peels of fruits etc., to cover the sand.

(iv) Use green leaves to cover over the sand.

(v) Sprinkle some water to make this layer wet.

(vi) Put some red worms on the upper layer of bed.

(vii) Feed vegetable and fruit wastes as food for them.

(viii) After 3 to 4 weeks we get loose soil like material in the pit.

(ix) Remove the material from the box, dry it in the sun.

(x) Use this as manure.

Advantages of Vermicomposting: **(S-17)**

1. This method increases moisture and nutrient retention of the soil.
2. It provides efficient conversion of organic wastes.
3. The method improves aeration and root penetration.
4. It reduces crusting of soil surface.
5. Micro-nutrients are added because of this method.
6. It increases the number of beneficial soil microorganisms.

Suitability of Vermicomposting: **(S-17)**

1. Vermicomposting plays a vital role in waste management. Hence, it is used in back of the house areas and kitchen.
2. It is eco-friendly technique, hence it can be adopted for organic waste from industries (viz. sugar, food industry etc.).
3. The nutrients from manure can be used for farms and gardens.
4. It improves the quality of fruits and vegetables.

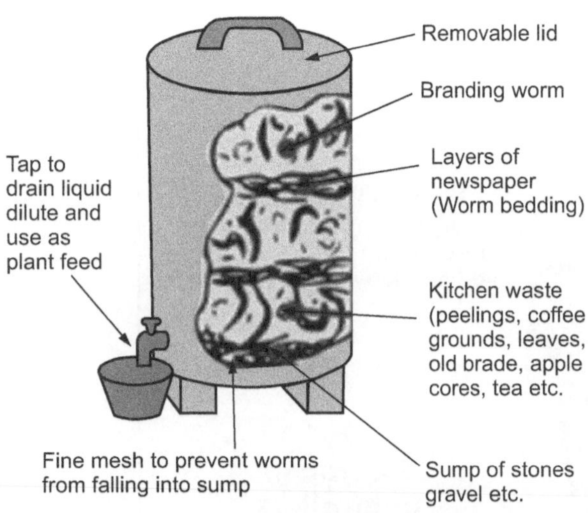

Fig. 3.11: Wormbin

Fig. 3.12: Worms used in Vermicompost

3.3 LANDFILLING TECHNIQUE

3.3.1 Definition

- The term 'landfill' is used herein to describe a unit operation for final disposal of 'Municipal Solid Waste' on land, designed and constructed with the objective of minimum impact to the environment.
- This term covers other terms such as 'secured landfill' and 'engineered landfills' which are also sometimes applied to municipal solid waste (MSW) disposal units.

- The term 'landfill' can be treated as synonymous to 'sanitary landfill' of Municipal Solid Waste, only if the latter is designed on the principle of waste containment and is characterised by the presence of a liner and leachate collection system to prevent ground water contamination. The term 'sanitary' landfill has been extensively used in the past to describe MSW disposal units constructed on the basis of 'dump and cover' but with no protection against ground water pollution. Protection against ground water pollution is therefore essential part of landfill.

3.3.2 Sorting Prior to Waste Processing or Land Filling (S-19)

- A SWM, depending on the level of complexity, will consist of a combination of unit processes in varying degrees of mechanization:
  1. **Pre-Sorting:** Bulky and contaminated wastes hamper further sorting/processing in the facility; mechanical or manual pre-sorting is essential to separate out these wastes.
  2. **Mechanical Sorting:** Mechanical processes based on principles of electro-magnetics, fluid mechanics, pneumatics etc. are used to segregate the different waste streams in the pre-sorted waste. Mechanical processes require specialized equipment for segregation of co-mingled municipal waste.
- Mechanical sorting typically employs the following processes:
  - **(a) Screening:** Screening segregates waste into two or more size distributions. Two types of screens are used in SWM centres: (1) Disc screens, (2) Trommels (Rotary screen).
  - **(b) Ferrous metal separation:** In the second stage, electromagnets are used for separating heavy ferrous metals from mixed waste.
  - **(c) Air classification:** The residual waste stream is passed through an air stream with sufficient velocity to separate light materials from heavy material.
  - **(d) Non-ferrous metal separation:** An eddy current segregates zinc, aluminum, copper, lead, nickel and other precious metal from commingled waste.
  - **(e) Detect and Route system:** This system separates out various grades of paper, plastics and glass, which are not sorted out in the air classifier.
  - **(f) Size reduction:** Sorted materials after segregation are usually too large for further use or processing, they should be reduced to smaller sizes.
  - **(g) Baling:** Sorted and sized material is baled for further processing/use.

3.3.3 Suitability of Waste for Landfilling

(A) Landfilling will be done for the following types of waste:

- (i) Comingled waste (mixed waste) not found suitable for waste processing;
- (ii) Pre-processing and post-processing rejects from waste processing sites;
- (iii) Non-hazardous waste not being processed or recycled.

(B) Landfilling will usually not be done for the following waste streams in the municipal solid waste:

- (i) Bio waste/garden waste;
- (ii) Dry recyclables.

(C) Landfilling of hazardous waste stream in the municipal waste will be done at a hazardous waste landfill site; such a site will be identified by the State Government and is likely to be operated by industries of a district/state. If such a landfill is not available, municipal authorities will dispose the hazardous waste in a special hazardous waste cell in the MSW landfill. Such a cell will be designed as per Ministry of Environment and Forests (MoEF) guidelines for hazardous waste disposal.

(D) Landfilling of construction and demolition waste will be done in a separate landfill where the waste can be stored and mined for future use in earthwork or road projects. If such a landfill site is not available, the waste will be stored in a special cell at a MSW landfill from where it can be mined for future use. Construction and demolition waste can be used as a daily cover at MSW landfills; however only minimum thickness of cover should be provided. All excess construction waste should be stored in the separate landfill cell.

(E) All existing and old landfills will be inspected and boreholes will be drilled for (i) recovery of leachate samples from the base of the landfill, (ii) recovery of subsoil samples beneath the base of the landfill for evaluation of permeability and soil properties and (iii) recovery of waste samples for waste characterisation. A minimum of 3 boreholes will be drilled with at least one borehole for each acre of landfill area.

The quality of leachate samples will be compared with (a) the ground water quality in existing bore wells 2 km away from the landfill and (b) the Central Pollution Control Board (CPCB) norms for limits of contaminants in leachate. If the leachate quality and the permeability of the subsoil strata is observed to be satisfactory, the existing landfill can continue to operate with bi-annual monitoring of leachate quality in the drilled boreholes.

(F) If the leachate quality is observed to be of poor quality with respect to the local ground water quality or with respect to the CPCB norms, steps will be taken to close the existing landfill site and remedial measures adopted. All future landfilling will be undertaken in properly designed and constructed new landfills.

(G) New landfills will be established as per the norms for siting, site investigations, design, construction and operation and closure.

(H) The estimated annual cost for setting up and operating new landfills as per norms given in this chapter is estimated to lie between ₹ 200 to 300 per tonne of waste generated (at 1998 prices, excluding land acquisition cost). Provisions may be made by the municipal authorities for allocating adequate financial resources for establishing new landfills.

3.3.4 Essential Components of Landfill

The seven essential components of a MSW landfill are:

(a) A liner system at the base and sides of the landfill which prevents migration of leachate or gas to the surrounding soil.

(b) A leachate collection and control facility which collects and extracts leachate from within and from the base of the landfill and then treats the leachate.

(c) A gas collection and control facility (optional for small landfills) which collects and extracts gas from within and from the top of the landfill and then treats it or uses it for energy recovery.

(d) A final cover system at the top of the landfill which enhances surface drainage, prevents infiltrating water and supports surface vegetation.

(e) A surface water drainage system which collects and removes all surface runoff from the landfill site.

(f) An environmental monitoring system which periodically collects and analyses air, surface water, soil-gas and ground water samples around the landfill site.

(g) A closure and post-closure plan which lists the steps that must be taken to close and secure a landfill site once the filling operation has been completed and the activities for long-term monitoring, operation and maintenance of the completed landfill.

3.3.5 Advantages and Disadvantages of Landfilling **(W-16, 18)**

Advantages of Landfilling:

(i) Volume can be increased with little addition of equipment.
(ii) Filled land can be reused for other community purposes.
(iii) Replanting the area with trees is possible.
(iv) Absorbs massive amounts of solid wastes.
(v) Low cost and ease of application, no high-technology.
(vi) Access to methane.

Disadvantages of Landfilling:

(i) Possibility of contamination of water sources by waste water resulting from landfill.

(ii) Requires proper planning, design and operation.

(iii) There is a leakage of air pollutant gases like methane, carbon dioxide.

(iv) Require large area.

(v) Difficult to operate in rainy season.

(vi) Produce leachate.

3.3.6 Factors for Site Selection

Factors affecting selection of site for land filling of solid waste: (S-15, 17; W-16)

1. Site should be easily approachable.
2. It should be located away from community area.
3. Sufficient quantity of soil should be available nearby site.
4. Waterlogged and flood prone areas should be avoided.
5. Local climate should be considered while selecting site for land filling.
6. Ground water should be very deep; it should not be less than 5 m.

3.3.6.1 Location

- A locational criteria may be specified by a regulatory agency (e.g. Pollution Control Board).
- In the absence of regulatory requirements, the following criteria are suggested:

(a) Lake or Pond: No landfill should be constructed within 200 m of any lake or pond. Because of concerns regarding runoff of waste water contact, a surface water monitoring program should be established if a landfill is sited less than 200 m from a lake or pond.

(b) River: No landfill should be constructed within 100 m of a navigable river or stream. The distance may be reduced in some instances for non-meandering rivers but a minimum of 30 m should be maintained in all cases.

(c) Flood Plain: No landfill should be constructed within a 100 year flood plain. A landfill may be built within the flood plains of secondary streams if an embankment is built along the stream side to avoid flooding of the area. However, landfills must not be built within the flood plains of major rivers unless properly designed protection embankments are constructed around the landfills.

(d) Highway: No landfill should be constructed within 200 m of the right of way of any state or national highway. This restriction is mainly for aesthetic reasons. A landfill may be built within the restricted distance, but no closer than 50 m, if trees and berms are used to screen the landfill site.

(e) Habitation: A landfill site should be at least 500 m from a notified habitat area. A zone of 500 m around a landfill boundary should be declared a No-Development Buffer Zone after the landfill location is finalised.

(f) Public parks: No landfill should be constructed within 300 m of a public park. A landfill may be constructed within the restricted distance if some kind of screening is used with a high fence around the landfill and a secured gate.

(g) Critical Habitat Area: No landfill should be constructed within critical habitat areas. A critical habitat area is defined as 'the area in which one or more endangered species live'. It is sometimes difficult to define a critical habitat area. If there is any doubt then the regulatory agency should be contacted.

(h) Wetlands: No landfill should be constructed within wetlands. It is often difficult to define a wetland area. Maps may be available for some wetlands, but in many cases such maps are absent or are incorrect. If there is any doubt, then the regulatory agency should be contacted.

(i) Ground Water Table: A landfill should not be constructed in areas where water table is less than 2 m below ground surface. Special design measures be adopted, if this cannot be adhered to.

(j) Airports: No landfill should be constructed within the limits prescribed by regulatory agencies (MOEF/CPCB/Aviation Authorities) from time to time.

(k) Water Supply Well: No landfill should be constructed within 500 m of any water supply well. It is strongly suggested that this locational restriction be abided by at least for down gradient wells. Permission from the regulatory agency may be needed if a landfill is to be sited within the restricted area.

(l) Coastal Regulation Zone: A landfill should not be sited in a coastal regulation zone.

(m) Unstable Zone: A landfill should not be located in potentially unstable zones such as landslide prone areas, fault zone etc.

(n) Buffer Zone: A landfill should have a buffer zone around it, upto a distance prescribed by regulatory agencies.

(o) Other criteria may be decided by the planners.

3.3.6.2 Search Area

- To identify the potential sites for a landfill a 'search area' has to be demarcated. The search area is usually governed by the economics of waste transportation. It is usually limited by the boundaries of the municipality.
- Typically search areas are demarcated on a map using a 'search radius' of 5 to 10 km, keeping the waste generating unit as the centre. Alternatively, the search area may be identified by adopting a range of 5 km all around the built-up city boundary. One should start with a small search area and enlarge it, if needed.

3.3.6.3 Development of a List of Potential Sites

- After demarcating the search area, as well as after studying the various restrictions listed in the locational critieria, areas having potential for site development should be identified. A road map may be used to show the potential sites that satisfy the locational criteria. Preliminary data collection should be undertaken with an aim of narrowing the list of sites to a few best-ranked sites.
- In areas where land availability is scarce, degraded sites such as abandoned quarry sites or old waste dump sites can be considered. Special design measures are required for such sites.

3.3.6.4 Data Collection

Several maps and other information need to be studied to collect data within the search radius. Some are discussed below.

(a) Topographic Maps: The topography of the area indicates low and high areas, natural surface water drainage pattern, streams and rivers. A topographic map will help to find sites that are not on natural surface water drains or flood plains. Topographical maps may be procured from Survey of India.

(b) Soil Maps: These maps, primarily meant for agricultural use, will show the types of soil near the surface. They are of limited use as they do not show types of soil a few metre below the surface. They can be procured from Indian Agricultural Research Institute (IARI).

(c) Land Use Plans: These plans are useful in delineating areas with definite zoning restrictions. There may be restrictions on the use of agricultural land or on the use of forest land for landfill purposes. These maps are used to delineate possible sites that are sufficiently away from localities and to satisfy zoning criteria within the search area. Such maps are available with the Town planning authority or the Municipality.

(d) **Transportation Maps:** These maps, which indicate roads and railways and locations of airports, are used to determine the transportation needs in developing a site.

(e) **Water Use Plans:** Such maps are usually not readily available. However, once potential areas are demarcated, the water use in those areas must be investigated. A plan indicating the following items should be developed: private and public tubewells indicating the capacity of each well, major and minor drinking water supply line(s), water intake wells located on surface water bodies and open wells.

(f) **Flood Plain Maps:** These maps are used to delineate areas that are within a 100 year flood plain. Landfill siting must be avoided within the flood plains of major rivers.

(g) **Geologic Maps:** These maps will indicate geologic features and bedrock levels. A general idea about soil type can be developed from a geologic map.

Such maps can be procured from Geological Survey of India. They may be used to identify predominantly sandy or clayey areas.

(h) **Aerial Photographs/Satellite Imagery:** Aerial photographs or satellite imageries may not exist for the entire search area. However such information may prove to be extremely helpful. Surface features such as small lakes, intermittent stream beds and current land use, which may not have been identified in earlier map searches, can be easily identified using aerial photographs.

(i) **Ground Water Maps:** Ground water contour maps are available in various regions, which indicate the depth of ground water below the land surface as well as regional ground water flow patterns. Such maps should be collected from Ground Water Boards or Minor Irrigation Tubewell Corporations.

(j) **Rainfall Data:** The monthly rainfall data for the region should be collected from the Indian Meteorological Department.

(k) **Wind Map:** The predominant wind direction and velocities should be collected from the Indian Meteorological Department.

(l) **Seismic Data:** The seismic activity of a region is an important input in the design of landfills. Seismic coefficients are earmarked for various seismic zones and these can be obtained from the relevant BIS code or from the Indian Meteorological Department.

3.3.6.5 Site Walk-Over and Establishment of Ground Truths

- A site reconnaissance will be conducted by a site walk-over as a part of the preliminary data collection. All features observed in various maps will be confirmed.
- Additional information pertaining to the following will be ascertained from nearby inhabitants: (a) Flooding during monsoons; (b) Soil type; (c) Depth to G.W. Table (as observed in open wells or tube wells); (d) Quality of groundwater and (e) Depth to bedrock.

3.3.6.6 Preliminary Boreholes and Geophysical Investigation

- At each site, as a part of preliminary data collection, one to two boreholes will be drilled and samples collected at every 1.5 m interval to a depth of 20 m below the ground surface. The following information will be obtained: (i) Soil type and stratification; (ii) Permeability of each strata; (iii) Strength and compressibility parameters (optional); (iv) Ground water level and quality and (v) Depth to bedrock.
- In addition to preliminary boreholes, geophysical investigations (electrical resistivity/seismic refraction/others) may be undertaken to assess the quality of bedrock at different sites.

3.3.6.7 Assessment of Public Reaction

- The public/nearby residents should be informed of the possibility of siting of a landfill in a nearby area as soon as a list of potential sites is developed. A preliminary assessment of public opinion regarding all the sites in the list is essential.

- A site may be technically and economically feasible yet may be opposed heavily by the public. The "not in my back yard" (NIMBY) sentiment is high initially. However, with proper discussion it can be overcome in some cases. Early assessment regarding how strong the NIMBY sentiment is, can significantly reduce the time and money spent on planning for a landfill site which may not materialise.
- In many instances residents around a proposed site co-operate if the landfill site owner's representative listens to concerns of the area residents and considers those concerns in designing and monitoring a site. Noise, dust, odour, increase in traffic volume and reduction in property value concern the area residents more than the fear of groundwater contamination.
- Public reaction is less hostile if landfilling is done in an area already degraded by earlier municipal waste dumps or other activities such as quarrying, ash disposal etc.

3.3.6.8 Selection of Few Best-Ranked Sites

From amongst a large number of sites, the selection of a final site will emerge from a two-stage approach.

(a) Selection of a few best-ranked sites (usually 2 sites, sometimes 3) on the basis of pathway and receptor related attributes.

(b) Selection of final site on the basis of environmental impact assessment, social acceptance and cost of disposal.

- For the selection of a few best ranked sites, the Ranking System based on Site Sensitivity Index developed by Ministry of Environment and Forests (MOEF) in 1991 is recommended.
- Only the following attributes should be considered in such a study as indicated: (a) Population within 1 km; (b) Distance to drinking water well/tubewell; (c) Use of sites by residents; (d) Distance to nearest offsite building; (e) Presence of airport; (f) Presence of roads; (g) Current land use; (h) Distance to critical habitat nearby; (i) Distance to nearest surface water; (j) Depth to ground water; (k) Soil permeability; (l) Depth to bedrock; (m) Susceptibility to flooding; (n) Susceptibility to water erosion; (o) Slope stability of final landform; (p) Air pollution potential and (q) Susceptibility to seismic activity.
- On the basis of the ranking, scores received by various sites, one or two sites (sometimes upto 3) may be chosen for environmental impact assessment and final selection.

3.3.6.9 Environmental Impact Assessment (EIA)

- Wherever feasible, environmental impact assessment will be conducted for two alternate sites (in exceptional circumstances upto 3 sites).
- The impact of the landfill on the following will be quantified: (a) Ground water quality; (b) Surface water quality; (c) Air quality – gases, dust, litter; (d) Aesthetics – visual, vermin, flies; (e) Noise; (f) Land use alteration; (g) Traffic alteration; (h) Drainage alteration; (i) Soil erosion; (j) Ecological impacts and (k) Others.
- A comparison of both alternatives amongst themselves as well as with the null alternative (that is what would happen if the project was not carried out) should be made and suitability of the sites is summarised.

3.3.6.10 Final Site Selection

- The final selection of the site from amongst the best-ranked alternatives should be done by comparing:

  (a) the environmental impact; (b) social acceptance; and

  (c) transportation and landfilling costs.
- Transportation costs may be compared on the basis of average hauling distance from the centre of the waste generating unit (city or part thereof).
- Landfilling costs are difficult to compute at the preliminary stage but may be compared on the basis of the shape of the completed landfill and material costs for liner system, leachate collection system, daily cover and final cover system.
- A landfill site with low environmental impact, high social acceptance and low costs is the most desirable. If conflicting results appear for (a), (b) and (c), environmental impact minimisation should normally be given top priority.

3.3.7 Land Filling Methods (or Land Filling Sections) (W-16; S-19)

3.3.7.1 Area Method (or Above Ground Landfilling) (W-15)

- The area method is best suited for flat or gently sloping areas where some land depressions may exist. Also the area landfill is used when the terrain is unsuitable for the excavation of trenches in which the solid waste is to be placed.
- High-groundwater conditions necessitate the use of area-type landfills. Site preparation includes the installation of a liner and leachate control system. Cover material must be hauled in by truck or earthmoving equipment from adjacent land or from borrow-pit areas.

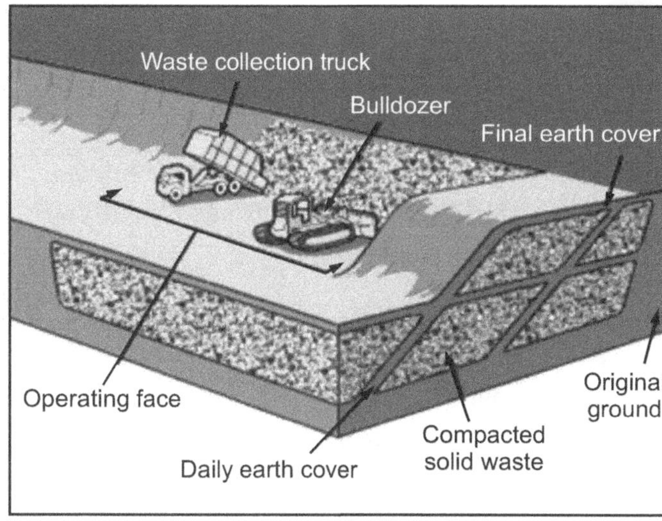

Fig. 3.13: Area Method

Advantages:

- Does not require excavation of trenches.
- Useful in areas where the terrain may be unsuitable for trench operations.
- Can accommodate high traffic volumes since the working face is not limited by the size of an excavation.

Disadvantages:

- Cover material may have to be imported.
- Potential exists for greater litter problems.
- Larger overall area/volume ratio (greater amount of landfill surface area used per volume of waste deposited) may result in higher costs.
- Topographic control not as obvious as for trench operation.

3.3.7.2 Trench Landfill (or Below Ground Landfilling)

- The trench method of landfilling is ideally suited to areas where an adequate depth of cover material is available at the site and where the water table is not near the surface. Typically, solid wastes are placed in trenches excavated in the soil. The soil excavated from the site is used for daily and final cover. The excavated trenches are lined with low-permeability liners to limit the movement of both landfill gases and leachate. Trenches vary from 100 to 300 m in length, 1 to 3 m in depth and 5 to 15 m in width with side slopes of 2 : 1.

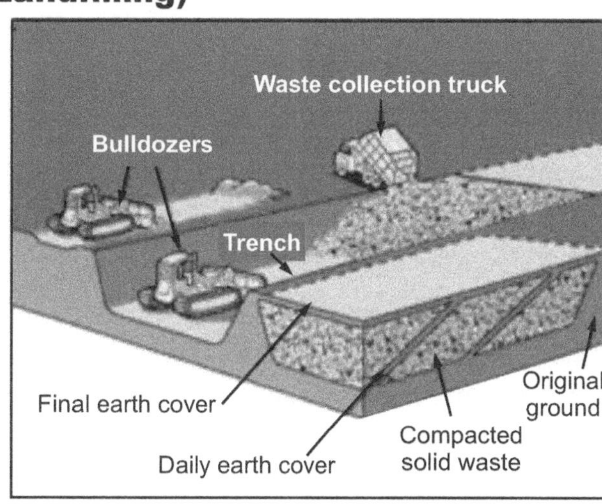

Fig. 3.14: Trench Method

Advantages:

- Not usually necessary to import cover material.
- Can be operated with a minimum working face exposed.
- Can be designed for optimum drainage during filling operations.
- May provide better litter control.
- Lower area/volume ratio (smaller amount of landfill surface area used per volume of waste deposited) may result in lower leachate production potential.

Disadvantages:

- Not suitable in geologic areas with high excavation costs or shallow ground water.
- Leachate collection may be more difficult/expensive.
- Loss of landfill volume due to the "between trench" areas.
- Trench depths and side slopes depend on soil types and stability.
- Trench size may be insufficient to handle.
- Unexpected high traffic volumes.

3.3.7.3 Slope Landfill

- In hilly regions it is usually not possible to find flat ground for landfilling. Slope landfills and valley landfills have to be adopted. In a slope landfill, waste is placed along the sides of existing hill slope.
- Control of inflowing water from hillside slopes is a critical factor in design of such landfills.
- The slope or ramp is sometimes used in combination with the other two methods. The wastes are spread on an existing slope, compacted and covered. This variation may be suitable for most areas. The cover materials usually come from just ahead of the working face.

3.3.7.4 Valley Landfill

- Depressions, low-lying areas, valleys, canyons, ravines, dry borrow pits etc. have been used for landfills. The techniques to place and compact solid wastes in such landfills change with the geometry of the site, the characteristics of the available cover material, the hydrology and geology of the site, the type of leachate and gas control facilities to be used and the access to the site. Control of surface drainage is often a critical factor in the development of canyon/depression sites.
- It is recommended that the landfill section be arrived at keeping in view the topography, depth to water table and availability of daily cover material.

3.3.8 Leachate and Its Control (S-16, 18)

3.3.8.1 Definition (S-15)

- Leachate can be defined as 'a liquid that is produced when water or another liquid comes in contact with solid waste'. Leachate is a contaminated liquid that contains a number of dissolved and suspended materials.

OR

- Leachate is any liquid that in passing through matter, extracts solutes, suspended solids or any other component of the material through which it has passed.
- Leachate is generated on account of the infiltration of water into landfills and its percolation through waste as well as by the squeezing of the waste due to self weight.

3.3.8.2 Leachate Quality

- The important factors which influence leachate quality include waste composition, elapsed time, temperature, moisture and available oxygen.
- In general, leachate quality of the same waste type may be different in landfills located in different climatic regions.
- Landfill operational practices also influence leachate quality.

Table 3.1: Typical Constituents of Leachate from MSW Landfills

Constituent		Range (mg/l)	
Type	**Parameter**	**Minimum**	**Maximum**
Physical	pH	3.7	8.9
	Turbidity	30 JTU	500 JTU
	Conductivity	480 m mho/cm	72500 m mho/cm
Inorganic	Total Suspended Solids	2	170900
	Total Dissolved Solids	725	55000
	Chloride	2	11375
	Sulphate	0	1850
	Hardness	300	225000
	Alkalinity	0	20350
	Total Kjeldahl Nitrogen	2	3320
	Sodium	2	6010
	Potassium	0	3200
	Calcium	3	3000
	Magnesium	4	1500
	Lead	0	17.2
	Copper	0	9.0
	Arsenic	0	70.2
	Mercury	0	3.0
	Cyanide	0	6.0
Organic	COD	50	99000
	TOC	0	45000
	Acetone	170	11000
	Benzene	2	410
	Toluene	2	1600
	Chloroform	2	1300
	Delta	0	5
	1,2 dichloroethane	0	11000
	Methyl ethyl ketone	110	28800
	Naphthalene	4	19
	Phenol	10	28800
	Vinyl Chloride	0	100
Biological	BOD	0	195000
	Total Coliform bacteria	0	100
	Fecal Coliform bacteria	0	10

3.3.8.3 Leachate Quantity

- The quantity of leachate generated in a landfill is strongly dependent on the quantity of infiltrating water.
- This, in turn, is dependent on weather and operational practices.
- The amount of rain falling on a landfill to a large extent controls the leachate quality generated. Precipitation depends on geographical location.

  Generation Rate in 'Active Area': The leachate generation during the operational phase from an active area of a landfill may be estimated in a simplified manner as follows:

  Leachate volume = (Volume of precipitation) + (Volume of pore squeeze liquid) – (Volume lost through evaporation) – (Volume of water absorbed by the waste).

  Generation Rate After Closure: After the construction of the final cover, only that water which can infiltrate through the final cover percolates through the waste and generates leachate. The major quantity of precipitation will be converted to surface runoff and the quantity of leachate generation can be estimated as follows:

  Leachate volume = (Volume of precipitation) – (Volume of surface runoff) – (Volume lost through evapotranspiration) – (Volume of water absorbed by waste and intermediate soil covers).

3.3.8.4 Effect of Leachate (S-15)

- Problems including clogging with mud or silt.
- Growth of micro-organisms in the conduit.
- The chemical composition of leachate can weaken pipe walls, which may then fail.
- The percolation of leachate will cause the soil pollution.
- Lechate will also contaminate the shallow ground water source in rainy season.
- It affect the DO of content water which is harmful for aquatic life.
- It will lead to the growth of waterborne diseases.

Leachate Management: (S-19)

Leachate management is done by:

1. Keep the waste as dry as possible, and do not introduce any liquid wastes to minimize leachate.
2. Leachate is carefully collected and shall be treated before its release on ground if possible.
3. Prevention of migration of leachate from landfill sides and landfill base to the sub-soil by a suitable liner system should be provided.
4. Using leachate treatment methods like natural system, biological treatment and Using Physicochemical treatment.

3.3.8.5 Leachate Control (S-15, 18; W-15, 18)

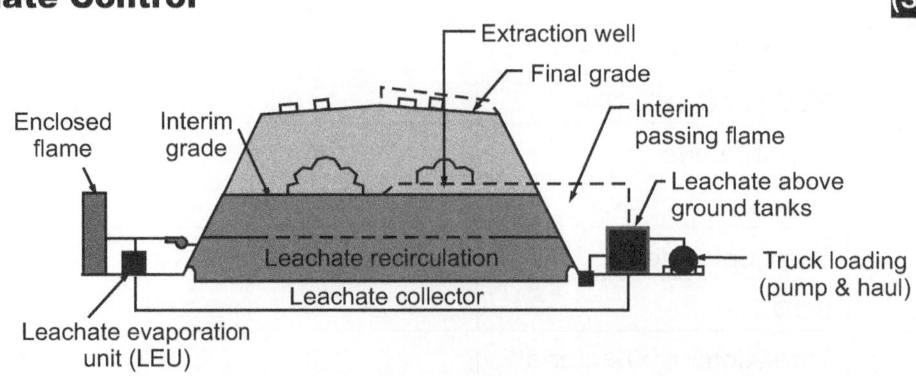

Fig. 3.15

- Leachate control within a landfill involves the following steps:

(a) Prevention of migration of leachate from landfill sides and landfill base to the subsoil by a suitable liner system; and

(b) Drainage of leachate collected at the base of a landfill to the sides of the landfill and removal of the leachate from within the landfill i.e. leachate is carefully collected and shall be treated before its release on ground if possible.

Liner systems are used for this purpose.

- Liner systems comprise of a combination of leachate drainage and collection layer(s) and barrier layer(s). A competent liner system should have low permeability, should be robust and durable and should be resistant to chemical attack, puncture and rupture.
- A liner system may comprise of a combination of barrier materials such as natural clays, amended soils and flexible geo-membranes.
- Use leachate treatment methods like Natural system, Biological treatment and Physicochemical treatment.
- Three types of liner systems are usually adopted and these are described hereafter:

(a) Single Liner System (S-18; W-15)**:** Such a system comprises of a single primary barrier overlain by a leachate collection system with an appropriate separation/ protection layer. A system of this type is used for a low vulnerability landfill.

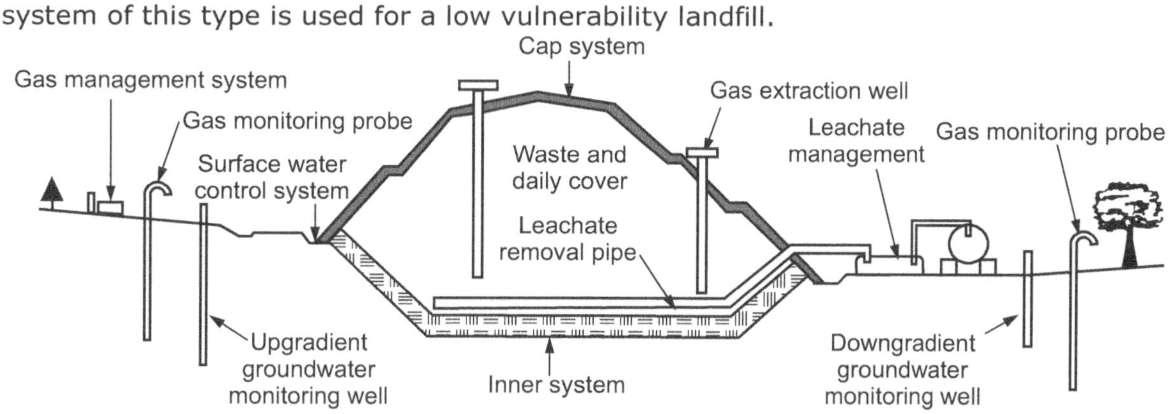

Fig. 3.16: Single Liner System

(b) Single Composite Liner System: A composite liner comprises of two barriers, made of different materials, placed in intimate contact with each other to provide a beneficial combined effect of both the barriers. Usually a flexible geo-membrane is placed over a clay or amended soil barrier. A leachate collection system is placed over the composite barrier. Single composite liner systems are often the minimum specified liner systems for non-hazardous wastes such as MSW.

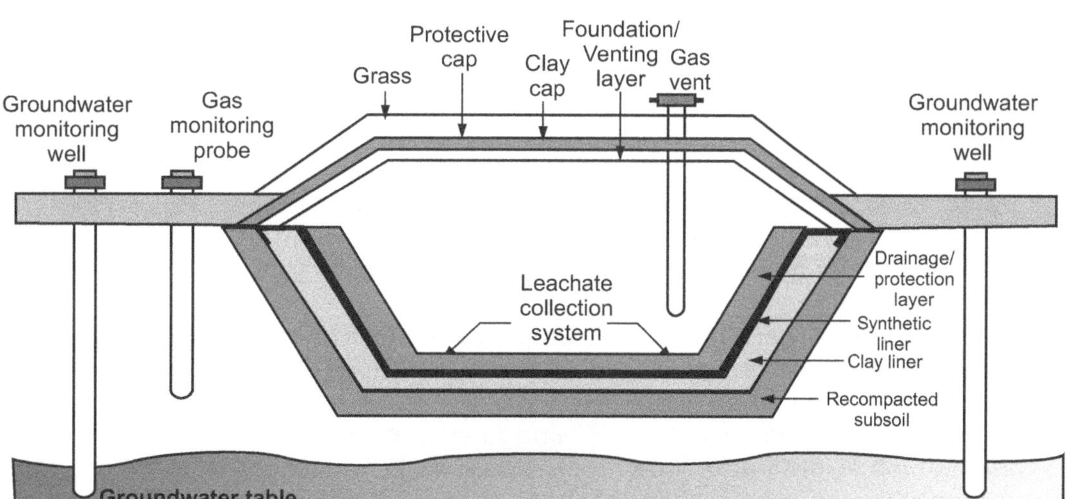

Fig. 3.17: Composite Liner System

(c) Double Liner System (S-18): In a double liner system a single liner system is placed twice, one beneath the other. The top barrier (called the primary barrier) is overlaid by a leachate collection system. Beneath the primary barrier, another leachate collection system (often called the 'leak detection layer') is placed followed by a second barrier (the secondary barrier). This type of system offers double safety and is often used beneath industrial waste landfills. It allows the monitoring of any seepage which may escape the primary barrier layer.

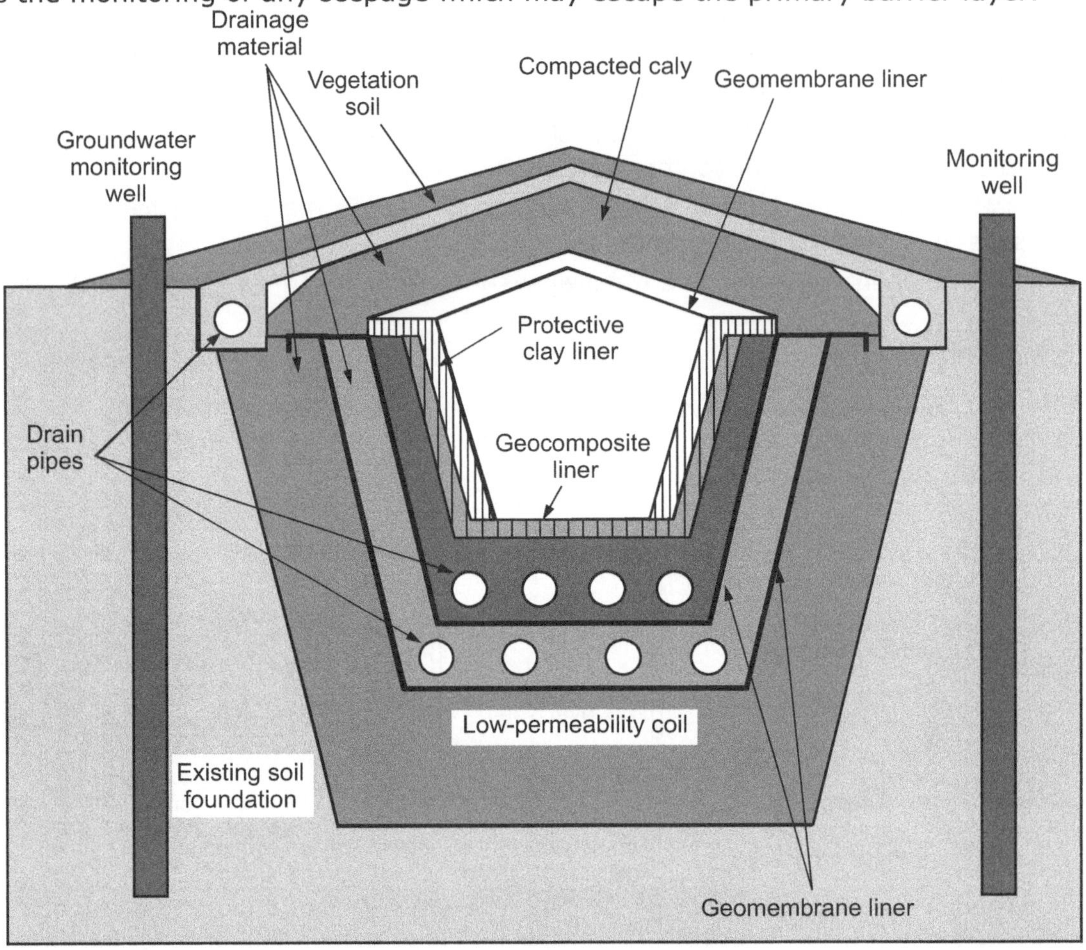

Fig. 3.18: Double Liner System

3.3.8.6 Leachate Drainage, Collection and Removal

- A leachate collection system comprises of a drainage layer, a perforated pipe collector system, sump collection area and a removal system.
- The leachate drainage layer is usually 30 cm thick, has a slope of 2% or higher and a permeability of greater than 0.01 cm/sec. A system of perforated pipes and sumps are provided within the drainage layer. The pipe spacing is governed by the requirement that the leachate head should not be greater than the drainage layer thickness. Pipe material selection is based on design requirements. HDPE pipes are most commonly used; other materials can also be examined for feasibility.
- Leachate is removed from the landfill by (a) pumping in vertical wells or chimneys, (b) pumping in side slope risers, or (c) by gravity drains through the base of a landfill in above ground and sloped landfills. Side slope risers are preferred to vertical wells to avoid any down drag problems.
- Submersible pumps have been used for pumping for several years; educator pumps are also being increasingly used. In some landfills, the leachate is stored in a holding tank (for a few days) before being sent for treatment.
- The possibility of fall in efficiency of the drainage system due to clogging associated with solid deposits and microbial growth is now well recognised. A number of options, including back flushing or breakthrough water after leachate head build-up need to be investigated at the design stage.

3.3.8.7 Leachate Management (S-19)

- The alternatives to be considered for leachate management are:
  - **(a) Discharge to Lined Drains:** This option is usually not feasible. It can only be adopted if the leachate quality is shown to satisfy all waste water discharge standards for lined drains, consistently for a period of several years.
  - **(b) Discharge to Waste Water Treatment System:** For landfills close to a waste water treatment plant, leachate may be sent to such a plant after some pre-treatment. Reduction in organic content is usually required as a pre-treatment.
  - **(c) Recirculation:** One of the methods for treatment of leachate is to recirculate it through the landfill. This has two beneficial effects: (i) the process of landfill stabilisation is accelerated and (ii) the constituents of the leachate are attenuated by the biological, chemical and physical changes occurring with the landfill. Recirculation of a leachate requires the design of a distribution system to ensure that the leachate passes uniformly throughout the entire waste. Since gas generation is faster in such a process, the landfill should be equipped with a well-designed gas recovery system.
  - **(d) Evaporation of Leachate:** One of the techniques used to manage leachate is to spray it in lined leachate ponds and allow the leachate to evaporate.

    Such ponds have to be covered with geo-membranes during the high rainfall periods. The leachate is exposed during the summer months to allow evaporation. Odour control has to be exercised at such ponds.
  - **(e) Treatment of Leachate:** The type of treatment facilities to be used depends upon the leachate characteristics. Typically, treatment may be required to reduce the concentration of the following prior to discharge: degradable and non-degradable organic materials, specific hazardous constituents, ammonia and nitrate ions, sulphides, odorous compounds and suspended solids. Treatment processes may be biological processes (such as activated sludge, aeration, nitrification (de-nitrification), chemical processes (such as oxidation, neutralisation) and physical processes (such as air stripping, activated adsorption, ultra-filtration etc.). The treated leachate may be discharged to surface water bodies. A leachate recirculation facility should be designed by a water supply specialist in conjunction with a geotechnical engineer.

3.3.9 Advantages and Disadvantages of Landfill (S-16)

Advantages	Disadvantages
1. Many a times landfill is a cheap waste disposal option for the local council.	1. The site will look ugly when it is being used for landfill.
2. Jobs will be created for local people.	2. Dangerous gases are given off from landfill sites that cause local air pollution and contribute to global warming.
3. Lots of different types of waste can be disposed off by landfill in comparison to other waste disposal methods.	3. Local streams could become polluted with toxins seeping through the ground from the landfill site.
4. The site could be re-landscaped and built-on, once it has been filled.	4. Once the site has been filled it might not be able to be used for redevelopment as it might be too polluted.
5. The gases given off by the landfill site could be collected and used for heating.	5. The jobs created in the local area are likely to be low paid jobs.
6. Waste transport costs will be reduced, as our waste will only have to travel a short distance to landfill. This will also reduce the amount of pollution caused by transporting the rubbish.	6. The trucks delivering the waste to the site are very big and noisy. They will spoil the peace and quiet in the area and could also contribute to traffic congestion (traffic jams) in the local area.

Advantages	Disadvantages
7. Local rubbish will be dealt locally instead of shifting the problem to another area or country e.g. Bangladesh.	7. The gases given off from the site as the waste begins to decompose (break down) will cause the air in the local area to be smelly.
8. Filled land can be reused for other community purposes.	8. It requires proper planning, design and operation.

3.3.10 Biogas from Landfill

Definition of Biogas:

Biogas typically refers to a mixture of different gases produced by the breakdown of organic matter in the absence of oxygen. Biogas can be produced from raw materials such as agricultural waste, manure, municipal waste, plant material, sewage, green waste or food waste.

- Landfill gas is generated as a product of waste biodegradation. Biological degradation of the waste may occur in the presence of oxygen (aerobic bacteria), in an environment devoid of oxygen (anaerobic bacteria), or with very little oxygen (facultative anaerobic bacteria).
- In all cases, organic waste is broken down by enzymes produced by bacteria in a manner comparable to food digestion. Considerable heat is generated by these reactions with methane, carbon dioxide and other gases as the by-products.
- Methane and carbon dioxide are the principal gases produced from the anaerobic decomposition of the biodegradable organic waste components in MSW. When methane is present in the air in concentrations between 5 and 15%, it is explosive. Because only limited amounts of oxygen are present in a landfill when methane concentrations reach this critical level, there is little danger that the landfill will explode.
- However, methane mixtures in the explosive range can form if landfill gas migrates off-site and mixes with air. The rate and quantity of gas generation with time, is difficult to predict. Typical generation rates reported in literature vary from 1.0 to 8.0 litres/kg/year. Landfill gas production rate of 6.0 cu.m. per hour has been reported from landfill sites in India having an area of 8 hectares and a depth of 5 to 8 m.
- The potential volume of landfill gas generation can be estimated to be 200 to 300 cu.m. per tonne of municipal waste. 50 to 75% of this gas can be recovered in mixed waste landfills using well functioning recovery systems. The recovery time is difficult to predict and may vary from 10 to 20 years or even more.

Table 3.2: Typical Constituents of Municipal Landfill Gas (W-18)

Constituents	Range
Methane	30 to 60%
Carbon Dioxide	34 to 60%
Nitrogen	1 to 21%
Oxygen	0.1 to 2%
Hydrogen Sulphide	0 to 1%
Carbon Monoxide	0 to 0.2%
Hydrogen	0 to 0.2%
Ammonia	0.1 to 1%

3.3.10.1 Working of Biogas Plant with the Help of Neat Sketch (S-17)

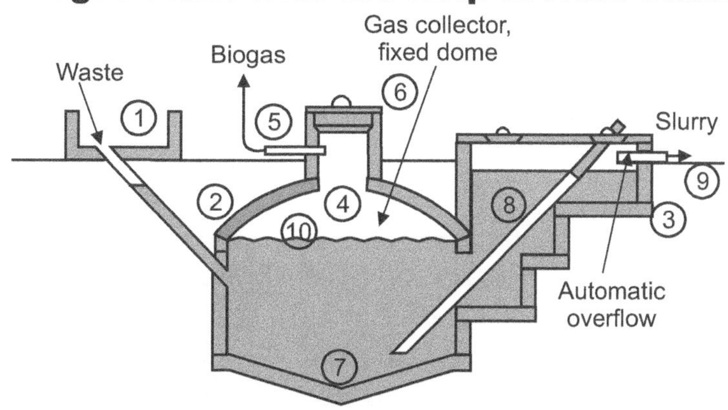

1. Mixing tank with inlet pipe and sand trap, 2. Digester, 3. Outlet chamber, 4. Gasholder, 5. Gas pipe, 6. Entry hatch with gastight seal, 7. Accumulation of thick sludge, 8. Outlet pipe, 9. Reference level, 10. Supernatant scum, broken up by varying level.

Fig. 3.19: Biogas Plant

1. The various forms of biomass are mixed with an equal quantity of water in the mixing tank. This forms the slurry.
2. The slurry is fed into the digester through the inlet chamber.
3. When the digester is partially filled with the slurry, the introduction of slurry is stopped and the plant is left unused for about two months.
4. During these two months, anaerobic bacteria present in the slurry decompose or ferment the biomass in the presence of water.
5. As a result of anaerobic fermentation, biogas is formed, which starts collecting in the dome of the digester.
6. As more and more biogas starts collecting, the pressure exerted by the biogas forces the spent slurry into the outlet chamber.
7. From the outlet chamber, the spent slurry overflows into the overflow tank.
8. The spent slurry is manually removed from the overflow tank and used as manure for plants.
9. The gas valve connected to a system of pipelines is opened when a supply of biogas is required.
10. To obtain a continuous supply of biogas, a functioning plant can be fed continuously with the prepared slurry.

3.4 INTRODUCTION AND PURPOSE OF RECYCLING (W-16)

3.4.1 Definitions of Recycling

- Recycling is the process of recovering and reusing waste product from household use, manufacturing, agriculture and business and thereby reducing their burden on the environment.
- Recycling is the process of collecting used materials, commonly known as waste and creating new products to prevent the waste of potentially useful materials.
- Recycling is the process of collecting and processing materials that would otherwise be thrown away as trash and turning them into new products.
- Recycling can benefit our community and the environment.

3.4.2 Purpose of Recycling (S-16; W-15, 16, 18)

- Less consumption of fresh raw materials to produce new products.
- Less energy consumption than that required to produce new products.
- Less air pollution (from incineration).
- Reducing the need for "conventional" waste disposal.
- Less water pollution (from landfilling).
- Lower greenhouse gas emissions.

3.4.3 Benefits of Recycling

(S-16, 18; W-15, 16)

- Reduces the amount of waste sent to landfills and incinerators.
- Conserves natural resources such as timber, water and minerals.
- Prevents pollution by reducing the need to collect new raw materials.
- Saves energy.
- Reduces greenhouse gas emissions that contribute to global climate change.
- Help sustain the environment for future generations.
- Helps create new well-paying jobs in the recycling and manufacturing industries in India.

Fig. 3.20: International Recycling Symbol

3.5 STEPS TO RECYCLING MATERIALS (S-15; W-16)

- Recycling includes the three steps below, which create a continuous loop, represented by the familiar recycling symbol.

Step-1: Collection and Processing

- There are several methods for collecting recyclables. There are four methods of collecting recyclables like
  - **(i) Curbside collection:** In curbside collection method, bins are placed at curbside for collection.
  - **(ii) Drop-off centers:** In drop-off centers, the procedures are required to carry recyclables to the centers.
  - **(iii) Buy back centers:** In buy back centers, the clean recyclables are purchased thus providing a clear incentive for use and creating a stable supply.
  - **(iv) Single stream collection:** In single stream collection, all recyclable materials are placed in a single bin for collection and sorting which is handled later at a central facility.
  - **(v) Deposit or refund programs:** After collection, recyclables are sent to a recovery facility to be sorted, cleaned and processed into materials that can be used in manufacturing. Recyclables are bought and sold just like raw materials would be and prices go up and down depending on supply and demand in India and the world.

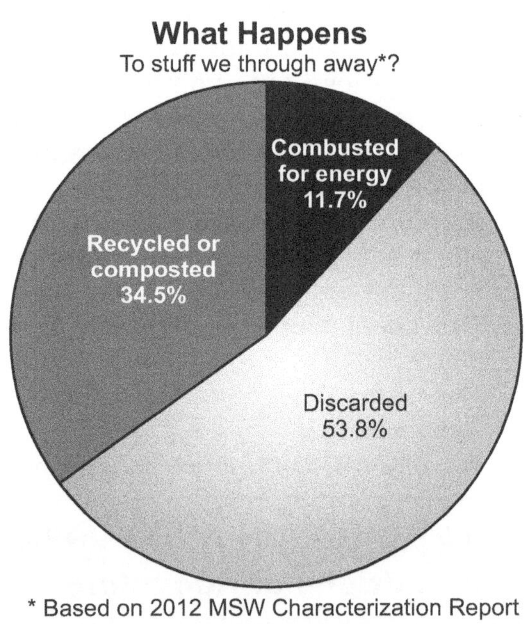

Fig. 3.21

Step-2: Sorting

- After collection of solid waste it is sorted out in different categories i.e. biodegradable and non-biodegradable material.
- Non-biodegradable material is also sorted out as plastic, metal, glass etc.

Step-3: Manufacturing

- More and more of today's products are being manufactured with recycled content.
- Common household items that contain recycled materials include:

- Newspapers and paper towels,
- Aluminium, plastic and glass soft drink containers,
- Steel cans, and
- Plastic laundry detergent bottles.
- Recycled materials are also used in new ways such as recovered glass in asphalt to pave roads or recovered plastic in carpeting and park benches.

Step-4: Purchasing New Products made from Recycled Materials

- By buying new products made from recycled materials you help close the recycling loop. There are thousands of products that contain recycled content.
- When you go shopping, look for:
  - Products that can be easily recycled and
  - Products that contain recycled content.

  Here are some of the terms used:
- **Recycled-content product:** This means, the product was manufactured with recycled materials, either collected from a recycling program or from waste recovered during the normal manufacturing process. Sometimes the label will tell you how much of the content was from recycled materials.
- **Post consumer content:** This is very similar to recycled content, but the material comes only from recyclables collected from consumers or businesses through a recycling program.
- **Recyclable product:** These are products that can be collected, processed and manufactured into new products after they have been used. These products do not necessarily contain recycled materials. Remember not all kinds of recyclables may be collected in your community.
- Some of the common products you can find that can be made with recycled content include:
  - Aluminium cans
  - Car bumpers
  - Carpeting
  - Cereal boxes
  - Comic books
  - Egg cartons
  - Glass containers
  - Laundry detergent bottles
  - Motor oil
  - Nails
  - Newspapers
  - Paper towels
  - Steel products
  - Trash bags

3.6 METHODS OF COLLECTION OF RECYCLABLES (S-16, 18; W-15, 18)

Recycling consists of a sequence of activities, such as collecting materials, sorting them and eventually processing recyclables into raw materials to manufacture new products. Recycling has several benefits since it helps to conserve the natural environment and decreases the quantity of solid waste disposed in landfills. The concept of recycling has been adopted across the world, necessitating collection of residential recyclables. Although collection of recyclable materials differs across communities, four distinct methods are widely used.

3.6.1 Curbside Collection

- Municipal recycling programs may provide communities with plastic bags containing source identification codes distinct to each community. These bags are packed with separated recyclable materials from different homes, sealed and left at the curb for collection by waste haulers.
- Systems of "Curbside Collection" encompass many different forms, which differ mostly on which point in the process the recyclables are sorted and cleaned. The different forms of curbside collection include "Mixed Waste Collection", "Commingled Recyclables" and "Source Separation".
- With "Mixed Waste Collection", the recyclables are mixed with the rest of the waste at collection. The recyclables would then need to be sorted out and cleaned at a central sorting facility. This results in a large amount of recyclables being wasted, especially paper which may be too soiled to reprocess. However, there is an advantage to this form of curbside collection – there is no need to pay for the separate collection of different recyclables and no public education is needed. Since all the sorting takes place in a central location, any change to which materials are recyclable can be easily accommodated.

- In a "Commingled Recyclable" system, the recyclables are separated from waste at collection. This greatly reduces the need for post-collection cleaning, but public education is required what materials are recyclable. Sorting is also required as different recyclables are mixed. Advances in sorting technology have lowered the cost of such a system, making it the preferred system currently.
- With "Source Separation", the recyclables are cleaned and sorted prior to collection. This method requires the least post-collection sorting and produces the purest recyclates. However, there may be additional operating costs to collect each type of recyclable separately. Extensive public education is also required, to minimize recyclate contamination in the process.

3.6.2 Drop-off Centers

- After separating the recyclables and putting them in special bags, individuals take these materials to centrally located collection sites such as charitable organizations, mobile collection stations or grocery stores. The materials are deposited in designated containers and taken to recycling plants. Drop centers are a preferable option of collecting recyclables in areas with sparse populations. This form of recycling waste collection is the easiest to establish. However, since the use of such centres is on a voluntary basis, it often suffers from low and unpredictable supply of recyclables.

3.6.3 Buy-back Centers

- Buy-back centers are facilities where individuals or communities take recyclables and trade them for payment. This incentive is meant to encourage more people to take their recyclables to centers and maintain a steady supply. The centers sell the post-processed materials to make some money and sustain the process. Government subsidies might make buy-back centers more viable, since the recycled materials might fetch a lower price than the total amount used in recycling.

3.6.4 Deposit Programs

- In these programs, customers pay an additional fee when purchasing beverage containers but receive their fee back once they return the container to the purchase point. Consumers can help to recycle as many beverage containers as possible to prevent these containers from ending up in the waste stream or as litter in the community. As an incentive, the Deposit Beverage Container Program places a certain amount as redeemable deposit on each beverage container. Consumers get back their amount when they return their containers to a redemption centre.

3.7 PRESENT STATUS OF RECYCLING SOLID WASTE IN INDIA (S-15, 18)

- In India, there is no formal recycling system but informal recyclers are there and play an important role in SWM. Recycling by means of repair, reprocessing, and reuse of waste materials is a common practice in India.
- These comprise of unorganized and unrecognized establishments which are difficult to be monitored by governmental agencies. However resource recovery through material recycling is taking place in India in a big way, through unorganized ways.
- Material recycling is done through sorting of waste into different streams at source or at a centralized facility. At present, recycling of dry recyclables does not take place at the household level in India.
- Waste is accessible to waste pickers; they segregate it into saleable materials such as paper, plastics, glasses, metal pieces, textile etc.
- Pickers segregate the wastes directly from the dumps and bins with no precautions and they are exposed directly to harmful wastes.
- The separated waste is sold to a small waste dealer, from where the waste is transferred to a medium sized dealer or wholesaler.
- All these activities are not regulated or monitored by any governmental organization. Due to this informal segregation, volume reduction is achieved, while it ignores social, economic, environmental and health aspects.

3.8 SOLID WASTE RECYCLING IN INDIA

- The rules in India require municipalities to ensure community participation in waste segregation (by not mixing "wet" food wastes with "dry" recyclables like paper, plastics, glass, metal etc.) and to promote recycling or reuse of segregated materials. In India, there is no formal recycling system but informal recyclers are there and play an important role in SWM. These comprise of unorganized and unrecognized establishments which are difficult to be monitored by governmental agencies. However resource recovery through material recycling is taking place in India in a big way, through unorganized ways.
- Material recycling is done through sorting of waste into different streams at the source or at a centralised facility. Sorting at source is more economical than sorting at a centralised facility.

3.8.1 Sorting at Source

- Sorting at source (home sorting) is driven by the existing markets for recyclable materials and the link between the house holder and the waste collector.
- The desirable home sorting streams are:
  (a) Dry recyclable materials e.g. glass, paper, plastics, cans etc.,
  (b) Bio-waste and garden waste,
  (c) Bulky waste,
  (d) Hazardous material in household waste,
  (e) Construction and Demolition waste and
  (f) Commingled MSW (mixed waste).
- At present, recycling of dry recyclables does take place at the household level in India. However, source separation and collection of waste in streams of (b), (c), (d) and (e) has to be developed in most cities.

3.8.2 Centralised Sorting

- Centralised sorting is needed wherever recyclable materials are collected in a commingled (mixed) state.
- Hand sorting from a raised picking belt is extensively adopted in several countries.
- Mechanised sorting facilities using magnetic and electric field separation, density separation, pneumatic separation, size separation and other techniques are used in some developed countries. Such facilities are usually very expensive in comparison to hand sorting.
- In India, centralised sorting is not adopted. However, some intermediate sorting does occur after household wastes reach curbside collection bins (dhalaos) through ragpickers. There is a need to formalise this intermediate sorting system or develop a centralised sorting facility to minimise recyclable materials reaching a waste processing facility or a landfill.

3.8.3 Sorting Prior to Waste Processing or Landfilling

- Home sorting and centralised sorting processes normally recover most of the recyclable materials for reuse. However, a small fraction of such materials may escape the sorting process. Sorting is also undertaken just prior to waste processing, waste transformation or landfilling to recover recyclable materials. In a landfill, sorting may be carried out by ragpickers immediately after spreading of a layer of waste. In waste processing or transformation centres, manual sorting or size separation is usually undertaken.
- Wherever manual sorting is adopted, care must be taken to ensure that sorters are protected from all disease pathways and work in hygienic conditions.

3.9 RESOURCE RECOVERY THROUGH WASTE PROCESSING (S-15, 19)

- Biological or thermal treatment of waste can result in recovery of useful products such as compost or energy.

3.9.1 Biological Processes (S-15)

- Biological treatment involves decomposition of the biodegradable components of waste using micro-organisms.

- Two types of processes are used, namely:
  (a) Aerobic processes: Windrow composting, aerated static pile composting and in-vessel composting; vermiculture etc.
  (b) Anaerobic processes: Low-solids anaerobic digestion (wet process), high-solids anaerobic digestion (dry process) and combined processes.
- In the aerobic process, the utilisable product is compost. In the anaerobic process, the utilisable product is methane gas (for energy recovery). Both processes have been used for waste processing in different countries – a majority of the biological treatment process adopted world-wide is aerobic composting; the use of anaerobic treatment has been more limited.
- In India, aerobic composting plants have been used to process upto 500 tons per day of waste.

3.9.2 Thermal Processes (S-15; W-15)

- Thermal treatment involves conversion of waste into gaseous, liquid and solid conversion products with concurrent or subsequent release of heat energy.
- Three types of systems can be adopted, namely:
  (a) Combustion systems (Incinerators): Thermal processing with excess amounts of air.
  (b) Pyrolysis systems: Thermal processing in complete absence of oxygen (low temperature).
  (c) Gasification systems: Thermal processing with less amount of air (high temperature).
- Combustion system is the most widely adopted thermal treatment process world-wide for MSW. Though pyrolysis is a widely used industrial process, the pyrolysis of municipal solid waste has not been very successful. Similarly, successful results with mass fired gasifiers have not been achieved. However, both pyrolysis and gasification can emerge as viable alternatives in the future.
- Three types of combustion systems have been extensively used for energy recovery in different countries namely: mass-fired combustion systems (MASS),
- Refuse Derived Fuel (RDF), fired combustion systems and Fluidised Bed (FB) combustion systems are mostly used.
- To be viable for energy recovery through thermal processing, the municipal solid waste must possess a relatively high calorific value. In the MSW generated in developed countries, presence of significant quantity of paper and plastics yields a high calorific value of the MSW (typically above 2000 kCal/kg) which makes it suitable for thermal processing.
- In Indian MSW, the near absence of paper and plastics as well as the presence of high quantities of inert material, all combine to yield a low calorific value of the MSW (typically less than 1000 kCal/kg). In its mixed form, such waste may not be suitable for thermal processing.
- However, removal of inerts from Indian MSW as well as development of combustion system for low-calorific value wastes can result in a reversal of this position in the future.

Recycling of (i) Fly ash, (ii) Blast furnace slags, (iii) Pulp and paper (iv) Chromium Sludge: (S-19)

(i) Fly ash: Fly ash waste from thermal plants is recycled for manufacturing of cement, light weight aggregate, insulating bricks, precast concrete, soil stabilization, land reclamation etc.

(ii) Blast furnace slags: Blast furnace slag from steel plants is recycled for manufacturing of cement, refractory, ceramic material, aggregates etc.

(iii) Pulp and paper: Pulp and paper from paper industry are recycled and again used for manufacturing new paper and paper products.

(iv) Chromium Sludge: Chromium sludge is recycled for manufacturing of ordinary cement, coloured cement etc.

3.9.3 Other Processes

- New biological and chemical processes which are being developed for resource recovery from MSW are:
  (a) Fluidised bed bio-reactors for cellulose production and ethanol production.
  (b) Hydrolysis processes to recover organic acids.
  (c) Chemical processes to recover oil, gas and cellulose.
  (d) Others.
- The economic viability of these processes is yet to be established.

Common Products that can be made with recycled content: (S-19)

The recycling of various industrial wastes such as Fly ash, Blast furnace slag, Lime sludge, Ferro alloy slag, Non-ferrous metal waste, Phosphogypsum, etc. are as follows.

1. Fly ash waste from thermal plants is recycled for manufacturing of cement, light weight aggregates, insulating bricks, precast concrete, soil stabilization, land reclamation etc.
2. Blast furnace slag from steel plants is recycled for manufacture of cement, refractory, ceramic material, aggregates etc.
3. Phosphogypsum waste from chemical plants is recycled for manufacture of gypsum plaster, boards, tiles, cement products etc.
4. Non-ferrous metal industry waste is recycled for manufacture of binder material, construction blocks, heavy clay products, coloured concrete, floor tiles, polymer doors etc.
5. Lime sludge from paper allied industries is recycled for manufacturing of building materials, lime, masonry cement, lime bricks, binder materials etc.

3.10 INCINERATION OF WASTE (W-15)

3.10.1 Definition

- It is the process of direct burning of wastes in the presence of excess air (oxygen) at temperatures of about 80°C and above, liberating heat energy, inert gases and ash.

3.10.2 Introduction

- Net energy yield depends upon the density and composition of the waste; relative percentage of moisture and inert materials, which add to the heat loss; ignition temperature; size and shape of the constituents; design of the combustion system (fixed bed/fluidised bed) etc. In practice, about 65 to 80% of the energy content of the organic matter can be recovered as heat energy, which can be utilised either for direct thermal applications, or for producing power via steam turbine generators (with typical conversion efficiency of about 30%).
- The combustion temperatures of conventional incinerators fuelled only by wastes are about 760°C in the furnace and in excess of 870°C in the secondary combustion chamber. These temperatures are needed to avoid odour from incomplete combustion but are insufficient to burn or even melt glass. To avoid the deficiencies of conventional incinerators, some modern incinerators utilise higher temperatures of upto 1650°C using supplementary fuel. These reduce waste volume by 97% and convert metal and glass to ash.
- Wastes burned solely for volume reduction may not need any auxiliary fuel except for start-up. When the objective is steam production, supplementary fuel may have to be used with the pulverized refuse, because of the variable energy content of the waste or in the event that the quantity of waste available is insufficient.
- While incineration is extensively used as an important method of waste disposal, it is associated with some polluting discharges which are of environmental concern, although in varying degrees of severity. These can fortunately be effectively controlled by installing suitable pollution control devices and by suitable furnace construction and control of the combustion process.

Process of Incineration: (S-17)

- Incineration is a chemical reaction in which carbon, hydrogen and other elements in the waste, mix with oxygen in the combustion zone and generates heat.

- The air requirements for combustion of solid wastes are considerable. Usually, excess air is supplied to the incinerator to ensure complete mixing and combustion and to regulate operating temperature and control emissions.
- Excess air requirements, however, differ with moisture content of waste, heating values and the type of combustion technology employed.
- Many incinerators are designed to operate in the combustion zone of 900°C – 1100°C. This temperature is selected to ensure good combustion, complete elimination of odours and protection of the walls of the incinerator.
- Incinerator systems are designed to maximise waste burn out and heat output, while minimising emissions by balancing the oxygen (air) and the three "Ts", i.e., time, temperature and turbulence.
- Complete incineration of solid wastes produces virtually an inert residue, which constitutes about 10% of the initial weight and perhaps a larger reduction in volume.
- The residue is generally landfilled. The incineration facility along with combustion of waste emits air pollutants (i.e., fine particulate and toxic gases), which are an environmental concern, and, therefore, their control is necessary.

3.10.3 Incinerators (S-16; W-16)

3.10.3.1 Definitions

1. Multiple Chamber Incinerators:

- It means an incinerator with two or more refractory-lined combustion chambers in series separated physically by refractory walls, interconnected by gas passages and employing adequate design parameters necessary for maximum combustion of the refuse materials.

2. Municipal Incinerators:

- It means a solid waste incinerator that combusts MSW exclusively.

Working of Municipal Incinerators: (W-15)

- Incineration is a waste treatment process that involves the combustion of solid waste at 1000°C.
- Waste materials are converted into ash, flue gas and heat.
- The ash is mostly formed by the inorganic constituents of the waste and gases due to organic waste.
- The heat generated by incineration is used to generate electric power.

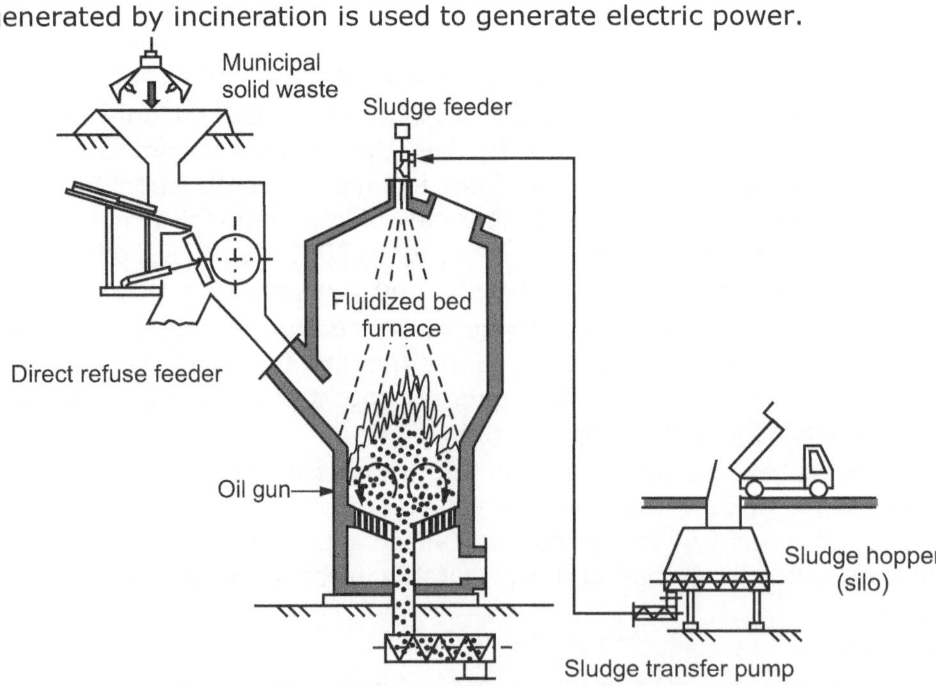

Fig. 3.22: Municipal Incinerator

3.10.3.2 Types of Incinerators (W-16, 18)

- Both Stoker and Fluidised Bed type furnaces are used in incinerators. The modern municipal incinerators are usually of the continuously burning type and may have "water wall" construction in the combustion chamber in place of the older, more common refractory lining. Corrosion of "water wall" units can, however, be a problem.
- Recent advancements include Twin Interchanging Fluidised Bed Combustor developed by a company in Japan, which is claimed to be capable of completely combusting wastes of low to high calorific values at very high overall efficiency.

Some basic types of Incineration Plants operating in the developed countries in the West and in Japan are as follows:

(i) Mass Burn Incinerator:

- Mass-burning systems are the predominant form of MSW incineration. A mass burn facility typically consists of a reciprocating grate combustion system and a refractory-lined, water-walled steam generator. Mass-burn systems generally consist of either two or three incineration units ranging in capacity from 50 to 1,000 tonnes per day. That is to say, the facility capacity ranges from about 100 - 150 to 2,000 - 3,000 tonnes per day. These facilities can accept/refuse that has undergone little preprocessing other than the removal of oversized items. Although this versatility makes mass burn facilities convenient and flexible, local programmes to separate household hazardous wastes (e.g., cleaners and pesticides) and recover certain materials (e.g., iron scrap) are necessary to help ensure environmentally viable incineration and resource conservation.
- About three-fourths of the waste-to-energy facilities in the U.S. and a few other countries are 'mass burn', where refuse is burned just as it is delivered to the plant, without processing or separation. These plants are sized to incinerate upto 3,000 tons of refuse per day and use two or more burners in a single plant. While facilities are sized according to the expected volume of waste, they are actually limited by the amount of heat produced when the garbage is burned. For example, if garbage burns hotter than it is expected to, less volume of material can be incinerated. Some mass burn plants remove metals from the ash for recycling. Mass burn plants have operated successfully in Europe for more than 100 years.
- At an MSW combustion facility, MSW is unloaded from collection trucks and placed in a trash storage bunker. An overhead crane is used to sort the waste and then lift it into a combustion chamber to be burned. The heat released from burning is used to convert water to steam. The steam is then sent to a turbine generator to produce electricity. The remaining ash is collected and taken to a landfill. Particulates are captured by a high-efficiency baghouse (a filtering system). As the gas stream travels through these filters, more than 99% of particulate matter is removed. Captured fly ash particles fall into hoppers (funnel-shaped receptacles) and are transported by an enclosed conveyor system to the ash discharger where they are wetted to prevent dust and mixed with the bottom ash from the grate. The ash residue is then conveyed to an enclosed building where it is loaded into covered, leak-proof trucks and taken to a landfill designed to protect against groundwater contamination. Ash residue from the furnace can be processed for removal of recyclable scrap metals.

(ii) Modular Combustion / Incinerator Units:

- Modular incinerators are simply small 'mass burn' plants with capacity ranging from 25 to 300 tonnes per day. The boilers are built in a factory and shipped to the plant site, rather than being erected on the site, as is the case with larger plants. These facilities are often used in small communities.
- Modular incinerator units are usually prefabricated units with relatively small capacities between 5 and 120 tonnes of solid waste per day. Typical facilities have between 1 and 4 units with a total plant capacity of about 15 to 400 tonnes per day. The majority of modular units produce steam as the sole energy product. Due to their small capacity, modular incinerators are generally used in small communities or for commercial and industrial operations. Their prefabricated design gives modular facilities, the advantage of a shorter construction time. Modular combustion systems are usually factory-assembled units consisting of a refractory-lined furnace and a waste heat boiler. Both units can be pre-assembled and shipped to the construction site, which minimises field installation time and cost. Adding modules or units,

installed in parallel can increase facility capacity. For example, a 200 tonne-per-day facility may consist of 4 units, a 50-tonne-per-day consists of 2 units and a 100 tonne-per-day consists of 1 unit. The number of units may depend on the fluctuation of waste generation for the service area and the anticipated maintenance cycle of the units.

(iii) Refuse-Derived Fuel (RDF) based Power Plants:

- In an RDF plant, waste is processed before burning. Typically, the non-combustible items are removed, separating glass and metals for recycling.
- The combustible waste is shredded into a smaller, more uniform particle size for burning. The RDF thus produced may be burned in boilers on-site, or it may be shipped to off-site boilers for energy conversion. If the RDF is to be used off-site, it is usually densified into pellets through the process of pelletisation.
- Pelletisation involves segregation of the incoming waste into high and low calorific value materials and shredding them separately, to nearly uniform size. The different heaps of the shredded waste are then mixed together in suitable proportion and then solidified to produce RDF pellets. The calorific value of RDF pellets can be around 4000 kcal/kg depending upon the percentage of organic matter in the waste, additives and binder materials used in the process, if any.
- Since pelletisation enriches the organic content of the waste through removal of inorganic materials and moisture, it can be very effective method for preparing an enriched fuel feed for other thermo-chemical processes like Pyrolysis/ Gasification, apart from Incineration.
- Additional advantage is that the pellets can be conveniently stored and transported.
- RDF plants involve significantly more sorting and handling than Mass Burn facilities and therefore provide greater opportunity to remove environmentally harmful materials from the incoming waste prior to combustion. However, it is not possible to remove the harmful materials completely. Several years ago RDF was used mainly along with coal fired boilers but now, because of the stricter restrictions w.r.t. air emissions, it is usually burned in dedicated boilers designed and built specially for the RDF. In case of RDF Pellets too, it needs to be ensured that the pellets are not burned indiscriminately or in the open, but only in dedicated incineration facilities or other well designed combustion systems, having all the necessary pollution control systems.

Refuse-derived Fuel (RDF) Incinerators:

- Refuse-derived fuel (RDF) refers to solid wastes in any form that is used as fuel. The term RDF, however, is commonly used to refer to solid waste that has been mechanically processed to produce a storable, transportable and more homogeneous fuel for combustion. RDF systems have two basic components: RDF production and RDF incineration.
- RDF production facilities make RDF in various forms through material separation, size reduction and pelletising. Although RDF processing has the advantage of removing recyclables and contaminants from the combustion stream, on an average, capital costs per tonne for incineration units that use RDF are higher than for other incineration options. RDF production plants like mass-burn incinerators characteristically have an indoor tipping floor. Instead of being pushed onto a pit, the waste in an RDF plant is typically fed onto a conveyor, which is either below grade or hopper fed. In some plants, the loader doing the feeding will separate corrugated and bulky items, like carpets.

(iv) Fluidised-bed incinerators

- Fluidised-bed incineration of MSW is typically medium scale, with processing capacity from 50 to 150 tonnes per day. In this system, a bed of limestone or sand that can withstand high temperatures, fed by an air distribution system, replaces the grate. The heating of the bed and an increase in the air velocities cause the bed to bubble, which gives rise to the term fluidised. There are two types of fluidised-bed technologies, viz., bubbling bed and circulating bed. The differences are reflected in the relationship between air flow and bed material, and have implications for the type of wastes that can be burned, as well as the heat transfer to the energy recovery system. Unlike mass-burn incinerators, fluidised-bed incinerators require front-end preprocessing, also called fuel preparation. They are generally associated with source separation because glass and metals do not fare well in these systems and also they can successfully burn wastes of widely varying moisture and heat content, so that the inclusion of paper and wood, which are both recyclable and burnable, is not a crucial factor in their operation (and thus paper can be extracted for higher-value recycling).

3.10.4 Use and Byproduct of Incineration Method of Solid Waste (S-15)

1. After the incineration process the left out products can be used as aggregate for preparation of low grade concrete or even sometimes it can also be used as road metal.
2. The incineration ash is used for making bricks or block manufacturing.
3. Also the steam generated during incineration can be used for electricity generation by running the turbines.
4. The products of incineration can also be used as filler material.

3.10.5 Products of Incineration Process with Their Use (S-16, 17, 19)

3.10.5.1 Heat

- The heat that is produced by an incinerator can be used for generating steam, which is used for driving a turbine in order to produce electricity. The typical amount as is produced by Municipal waste per ton is 2/3 MWh for electricity and 2 MWh for heating.

3.10.5.2 Solidified Outputs (S-16)

- Flue ash and Bottom ash is produced with the processing of waste materials and settle at the bottom of the incineration plant. The ash, which is produced, is 4 to 5% of total weight of the waste processed while the flue ash makes up some 10 to 20% of total weight of waste material.
- The heavy metals, which are contained in the flue or bottom ash, are lead, cadmium, zinc and copper. A small amount of furans and dioxins are also produced. It is to mention here that bottom ash seldom have heavy metals in it. Flue ash is hazardous while bottom ash is not that dangerous or injurious to health and that can be used as construction material since it is having cementing properties and also it can be used to make bricks.

3.10.5.3 Furans and Dioxins

- The biggest most concern, which has caught thoughts of environmentalists about MSW's incineration, is production of a huge amount of furans and dioxins. These are considered staidly injurious to health. Modern generators are equipped with special equipment to clean emission of gases from these injurious components.
- There was a time when no governmental regulations were there to bound incineration and save environment and atmosphere from this hazardous emission of gases but today there are strict and rigid rules and regulations to follow and conduct incineration. As such hence, they are harmful rather than useful.

3.10.5.4 Carbon Dioxide

- Incineration while being conducted produces a vast amount of carbon dioxide. Carbon dioxide plays a due role in global warming, as this is the green house gas. It has been observed that almost every thing which has carbon in its composition is when processed by incineration evolves out as carbon dioxide. As such hence it is harmful rather than useful.

3.10.5.5 Extra Emissions

- Some other emissions of gases by waste processing are sulfur dioxide, hydrochloric acid, fine particles and heavy metals which are not at all useful. Rather these are harmful for environment and are to be released only after cleaning the flue gas out of these.

Other Products of Incineration Process:

1. Residual ash
2. Fly ash
3. Trace organic and inorganic compounds
4. Waste gases

3.10.6 Advantages and Disadvantages of Incineration Process

3.10.6.1 Advantages (W-15, 18; S-19)

- The first concern for incineration stands against its injurious effects over health due to production of furans and dioxin emission. However, the emission is controlled to greater extent by developing of modern plants and governmental regulations.
- Incineration plants are capable of producing energy and can substitute power generation plants of other sort.
- The bottom ash after the process is completed, is considered non-injurious, that still is capable for being land filled and recycled.
- Fine particles are removable by processing through filters and scrubbers.
- Treating and processing medical and sewage waste produces non-injurious ash as product.
- Requires minimum land.
- Can be operated in any weather.
- Produces stable odour-free residue.
- Refuse volume is reduced by half.

3.10.6.2 Disadvantages (W-15)

- Extremely injurious matter needs adequate disposing off. This requires additional miles and need special locations for land filling this material.
- Although after a lot of regulations and restrictions and developments, concerns are still alive about emission of furans and dioxins.
- Incinerating plants are producers of heavy metals, which are injurious even in minor amounts.
- Solidified output from plant consists of a considerably high level of heavy metals and can prove fatal if they are not disposed off or reused properly.
- Initial investment costs are only recovered through long periods of contract for incinerating plants.
- Local communities always have opposed the presence of incinerating plant in the locality.
- The upheld view is to recycle, reuse and waste reduction instead of incineration.
- Expensive to build and operate.
- High energy requirement.
- Requires skilled personnel and continuous maintenance.
- Unsightly - smell, waste, vermin.

3.11 PYROLYSIS OF WASTE

3.11.1 Definitions (S-15, 16, 19; W-15, 16)

- **Pyrolysis** is also referred to as destructive distillation or carbonization. It is the process of thermal decomposition of organic matter at high temperature (about 900°C) in an inert (oxygen deficient) atmosphere or vacuum, producing a mixture of combustible Carbon Monoxide, Methane, Hydrogen, Ethane i.e. [CO, CH_4, H_2, C_2H_6] and non-combustible Carbon Dioxide, Water, Nitrogen i.e. [CO_2, H_2O, N_2] gases, pyroligenous liquid, chemicals and charcoal.
- The pyroligenous liquid has high heat value and is a feasible substitute of industrial fuel oil. Amount of each end-product depends on the chemical composition of the organic matter and operating conditions. Quantity and chemical composition of each product changes with pyrolysis temperature, residence time, pressure, feed stock and other variables.
- **Gasification** involves thermal decomposition of organic matter at high temperatures in presence of limited amounts of air/oxygen, producing mainly a mixture of combustible and non-combustible gas (Carbon Monoxide, Hydrogen and Carbon Dioxide).
- This process is similar to Pyrolysis, involving some secondary/different high temperature (>1000°C) chemistry which improves the heating value of gaseous output and increases the gaseous yield (mainly combustible gases $CO + H_2$) and lesser quantity of other residues. The gas can be cooled, cleaned and then utilized in IC engines to generate electricity.

- Pyrolysis/Gasification is already a proven method for homogenous organic matter like wood, pulp etc. and is now being recognised as an attractive option for MSW also. In these processes, besides net energy recovery, proper destruction of the waste is also ensured. The products are easy to store and handle. These processes are therefore being increasingly favoured in place of incineration.

3.11.2 Principle of Pyrolysis (S-15)

- Pyrolysis is the one of the most common methods in thermal conversion technology of biomass. In pyrolysis, biomass is heated to moderate temperatures, 400-600°C, in the absence of oxygen to produce oil. In gasification, biomass is heated to high temperatures, >700°C, to produce a synthesis gas (H_2 and CO), which can be converted in a catalytic step to liquid transportation fuels.

3.11.3 Methods of Pyrolysis (W-16; S-19)

Following methods of pyrolysis

(i) Garrets flash pyrolysis.

(ii) Pyrolysis developed by Energy research center of Bureau of Mines Pittsburg.

(iii) Destrugas gasification system. (iv) Slurry carb process.

(v) Plasma pyrolysis. (vi) Dry pyrolysis (vi) Oxidizing pyrolysis

(i) Garrets Flash Pyrolysis Process:

- This low temperature pyrolysis process has been developed by Garrett Research and Development Company.
- In a 4 tonnes per day pilot plant set up by the company at La Varne, California, the solid waste is initially coarse shredded to less than 50 mm size, air classified to separate organics/inerts and dried through an air drier.
- The organic portion is then screened, passed through a hammer mill to reduce the particle size to less than 3 mm and then pyrolysed in a reactor at atmospheric pressure.
- The branded heat exchange system enables pyrolytic conversion of the solid waste to a viscous oil at 500°C.

(ii) Pyrolysis Process developed by Energy Research Centre of Bureau of Mines, Pittsburg:

- This is a high temperature pyrolysis process to produce both fuel oil and fuel gas and has been investigated mainly at laboratory scale.
- The waste charge is heated in a furnace with nickel-chromium resistors to the desired temperature. The produced gases are cooled in an air trap where tar and heavy oil condense out.
- Uncondensed vapours pass through a series of water-cooled condensers where additional oil and aqueous liquors are condensed.
- The gases are then scrubbed in an electrostatic precipitator before further use. It is claimed that one tonne of dried solid waste produces 300-500 m^3 of gas, but the process is yet to be tested at full scale.

(iii) Destrugas Gasification System:

- In this system, the raw solid waste is first subjected to shredding /size reduction in an enclosed shed.
- The air from this shed is taken up as intake air in the plant so as to avoid odour problems.
- The shredded waste is fed to retorts (heated indirectly by burning gas in a chamber enveloping it) through which it sinks under gravity and gets subjected to thermal decomposition.
- The produced gas is washed and most of it (85%) used for heating the retorts. The remaining 15% is available as fuel.

(iv) Slurry Carb Process:

- This process has been developed by a company in USA to convert municipal solid waste into fuel oil. It is used in conjunction with a wet resource recovery process to separate out the recyclables.
- The received waste is first shredded and placed in an industrial pulper. The heavier and denser inorganic material sink to the bottom of the water-filled pulper from where it is easily removed.

- The remaining waste slurry (organic fraction) is subjected to violent pulping action, which further reduces the size of its constituents.
- The pulped organic waste is then subjected to high pressure and temperature whereby it undergoes thermal decomposition/carbonisation (slow pyrolysis) to fuel oil.

(v) Plasma Pyrolysis Vitrification (PPV)/Plasma Arc Process:

- This is an emerging technology utilising thermal decomposition of organic wastes for energy /resource recovery.
- The system basically uses a Plasma Reactor which houses one or more Plasma Arc Torches which generate, by application of high voltage between two electrodes, a high voltage discharge and consequently an extremely high temperature environment between 5000 - 14,000°C.
- This hot plasma zone dissociates the molecules in any organic material into the individual elemental atoms while all the inorganic materials are simultaneously melted into molten lava.
- The waste material is directly loaded into vacuum in a holding tank, preheated and fitted to a furnace where the volatile matter is gasified and fed directly into the plasma arc generator where it is pre-heated electrically and then passed through the plasma arc dissociating it into elemental stages.
- The gas output after scrubbing comprises mainly of CO and H_2. The liquefied produce is mainly methanol.
- The entire process is claimed to safely treat any type of hazardous or non-hazardous materials. It has the advantage that the NO_X (Oxides of Nitrogen) and SO_X (Oxides of Sulphur) gases emissions do not occur in normal operation due to the lack of oxygen in the system.
- Some US companies offering PPV technology are reported to be setting up some demonstration units based on this technology in Malaysia and Singapore.

(vi) Dry pyrolysis:

- Dry pyrolysis is the process of thermal decomposition without access of oxygen (O_2).
- Products of dry pyrolysis are gas with high heat of combustion, liquid and solid carbon residue.
- Depending on the temperature of the process there are three types of dry pyrolysis process. i.e. low temperature analysis, medium temperature analysis, high temperature analysis.

(vii) Oxidizing pyrolysis:

- It is impossible to achieve a completely oxygen-free atmosphere.
- Thus, a small amount of oxidation occurs in the process. If volatile or semi-volatile materials are present in the waste, thermal desorption will also occur.
- Thermal decomposition of industrial waste will occur by its partial burning or direct contact with end product of fuel combustion.
- This method is used for neutralization of most wastes including "inconvenient" ones for burning.

3.11.4 Merits and Demerits of Pyrolysis

Merits of Pyrolysis:

- Reduces greenhouse gas emissions and waste going to landfill.
- Produces a marketable product like gases, a bio-oil, bio-chemicals and charcoal.
- Low risk of water pollution.
- Low risk of odours.
- High recovery rate of resources.
- Minimal risk of health consequences.
- Commercially proven technology.

Demerits of Pyrolysis:

- Less capacity.
- Less efficiency.
- More expensive.
- Technology is still evolving.
- Markets are yet to be developed for char product and pyrolysis liquids.

3.12 GASIFICATION (S-19)

- Gasification is a process that converts organic or fossil fuel-based carbonaceous materials into carbon monoxide, hydrogen and carbon dioxide. This is achieved by reacting the material at high temperatures (>700°C), without combustion, with a controlled amount of oxygen and/or steam.

OR

- Gasification is the thermal decomposition of organic matter at high temperature in presence of limited air/oxygen, producing mainly a mixture of combustible and noncombustible gas (CO, H_2 and CO_2).

Important Points

- Composting is an organised method of producing compost manure.
- This manure contains plant nutrients like nitrogen, phosphorous and potassium.
- Composting can be carried out aerobically and anaerobically.
- Organisms, moisture, temperature, aeration, use of cultures, C/N ratio, addition of sewage and sewage sludge are the factors which affect the composting process.
- Banglore method (anaerobic) and Indore method (aerobic) are the manual methods of composting.
- To produce a manure on a large scale, mechanical composting is useful.
- Vermicompost is the process of composting using various worms, to create a heterogeneous mixture of decomposing vegetable or food waste.
- Landfill is unit operation for final disposal of MSW on land.
- Area method, Trench landfill, Slope landfill, Valley landfill are the methods of landfilling.
- Leachate is a liquid that is produced when water comes in contact with solid waste.
- Landfill gas is generated as a product of waste biodegradation.
- The process of turning of waste is called as incineration.
- The process of thermal decomposition of organic matter at high temperature (about 900°C) in vacuum is called as pyrolysis.
- Recycling is the process of recovering and reusing waste product.
- Collecting and processing, manufacturing and purchasing new products made from recycled materials are the three steps for recycling the materials.
- Curbside collection, drop-off centers, buy-back centers, deposit programs are the methods of collection of recyclables.
- Biological or thermal treatment of waste can result in recovery of useful products, such as compost or energy.
- In biological process, aerobic and anaerobic bacteria are used.
- Combustion systems, Pyrolysis systems and Gasification systems are used in thermal process.

Practice Questions

1. Write down the advantages and disadvantages of incineration process.
2. Describe in brief the methods of pyrolysis.
3. Define: (i) Pyrolysis; (ii) Gasification
4. State products of incineration process with their use
5. State the types of incinerators and describe each in brief.
6. Define: (i) Municipal Incinerators; (ii) Multiple Chamber Incinerators
7. Define and describe in brief 'the incineration of waste'.
8. State the advantages and disadvantages of landfill.
9. Give the typical constituents of municipal landfill gas.
10. How is the leachate management done?
11. Describe how is the leachate drained, collected and removed.

12. What is leachate control? What are Single Liner System, Single Composite Liner System and Double Liner System? Draw the neat labelled sketches of each.
13. How is the leachate quantity calculated?
14. What are the factors that affect the leachate quality? Give the typical composition of leachate.
15. Define leachate and state the reason for generation of it.
16. Describe the area method of landfilling. Write its advantages and disadvantages.
17. Describe the following:
    (i) Trench landfilling; (ii) Valley landfilling; (iii) Slope landfilling
18. State the factors affecting the site selection for landfilling.
19. What are the locational criteria for the site selection for landfilling?
20. What are the types of data that are to be collected for a site of landfilling?
21. How is the Assessment of Public Reaction done for siting of landfilling?
22. How is the Environmental Impact Assessment (EIA) done for landfilling site selection.
23. How is the Final Site Selection done for landfilling site selection.
24. What are the essential components of landfill?
25. How is the suitability of waste for landfilling decided.
26. Define landfill.
27. What is vermicomposting? Describe its procedure.
28. Describe mechanical composting.
29. Describe Indore and Bangalore methods of composting. What is the main difference between these?
30. Write down the factors affecting the composting process?
31. Describe the principles of composting.
32. What is: (i) Composting of waste and (ii) Disposal of waste.
33. Enlist methods of collection of recyclables.
34. How is the resource recovery through waste processing done?
35. Explain the Biological Processes for resource recovery through waste processing.
36. Explain the Thermal Processes for resource recovery through waste processing.
37. Explain sorting prior to waste processing or landfilling.
38. What are the other processes for resource recovery through waste processing?
39. What are the desirable home sorting streams?
40. What is central sorting in SW recycling? Explain
41. Explain sorting at source in SW recycling?
42. Describe solid waste recycling in India.
43. What are the methods of collection of recyclables? Explain.
44. Enlist the common products you can find that can be made with recycled content.
45. Explain the steps to recycling materials.
46. What are the benefits of recycling? Explain.
47. What is the purpose of recycling? Explain.
48. Define recycling.
49. What are the buyback centres?

MSBTE Questions and Answers (As Per 'G' Scheme)

Summer 2016

1. Describe the process of Indoor method of composting. **(4M)**

Ans. Refer Section 3.2.1.2.

2. State the products of incineration process with their use. **(4M)**

Ans. Refer Section 3.10.5.

3. Describe the process of disposal of solid waste by trench method of land filling. **(4M)**

Ans. Refer Section 3.3.7.2.

4. State the factors affecting on composting process of solid waste. **(4M)**

Ans. Refer Section 3.1.2.

5. State the four advantages and four disadvantages of Landfill method. **(4M)**

Ans. Refer Section 3.3.9.

6. What do you mean by Pyrolysis of waste and state its methods. **(4M)**

Ans. Refer Section 3.11.1.

7. What is vermicomposting. State its concept. **(4M)**

Ans. Refer Section 3.2.3.

8. What is leachate? How it can be controlled? **(4M)**

Ans. Refer Section 3.3.8.

9. Explain the types of incinerators. **(4M)**

Ans. Refer Section 3.10.3.2.

10. Describe the methods of collecting recyclables. **(4M)**

Ans. Refer Section 3.6.

11. State the purpose of recycling. **(4M)**

Ans. Refer Section 3.4.2.

12. State the factors affecting on composting process of solid waste. **(4M)**

Ans. Refer Section 3.1.2.

Winter 2016

1. State various factors affecting composting process. **(4M)**

Ans. Refer Section 3.1.2.

2. State the purpose of recycling of solid waste. **(4M)**

Ans. Refer Section 3.4.2.

3. Explain the principle of composting process. **(4M)**

Ans. Refer Section 3.1.1.

4. Enlist any four factors affecting site selection for landfilling. **(4M)**

Ans. Refer Section 3.3.6.

5. Enlist various methods of land filling and explain any one of them. Give the advantages and disadvantages of land filling. **(8M)**

Ans. Refer Section 3.3.7 and 3.3.5 respectively.

6. Enlist types of incinerators. Explain any one of them. Give the advantages and disadvantages of incineration process. **(8M)**

Ans. Refer Sections 3.10.3.2 and 3.10.6.

7. What do you understand by Vermi-composting? **(8M)**

Ans. Refer Section 3.2.3.

8. Explain pyrolysis of waste. **(4M)**

Ans. Refer Sections 3.11.1 and 3.11.3.

9. Describe the process of recycling. **(4M)**

Ans. Refer Sections 3.4.1.

10. State the purpose of recycling of solid waste. **(4M)**

Ans. Refer Section 3.4.2.

Summer 2017

1. State the four disadvantages of landfill method. **(4M)**

Ans. Refer Section 3.3.5.

2. State the four benefits of recycling of Solid Waste. **(4M)**

Ans. Refer Section 3.4.3.

3. Explain principle of composting process and favourable conditions for it. **(4M)**

Ans. Refer Section 3.2.

4. Differentiate between Bangalore and Indore methods of composting. **(4M)**

Ans. Refer Section 3.1.2.3.

5. Explain advantages of vermicomposting method also mention its suitability. **(4M)**

Ans. Refer Section 3.2.3.

6. State the four factors affecting the site selection for landfilling technique to dispose solid waste. **(4M)**

Ans. Refer Section 3.3.6.10.

7. Explain the mechanical composting method. **(4M)**

Ans. Refer Section 3.2.2.

8. Explain the working of biogas plant with the help of neat sketch. **(4M)**

Ans. Refer Section 3.3.10.1.

9. Describe the process of incineration and products of it with their uses. **(8M)**

Ans. Refer Sections 3.10.2 and 3.10.5.

10. State the benefits of recycling and marketing strategies for recyclables. **(8M)**

Ans. Benefits of Recycling: Refer Sections 3.4.3 and 3.5.

Winter 2017

1. List various transportation equipments in solid waste management system. **(2M)**

Ans. Refer Section 3.3.

2. State principle of vermi composting. **(2M)**

Ans. Refer Section 3.2.3.

3. State principle of composting. **(2M)**

Ans. Refer Section 3.1.1.

4. State any two benefits of recycling of solid waste. **(2M)**

Ans. Refer Section 3.4.3.

5. State any two steps in recycling of solid waste. **(2M)**

Ans. Refer Section 3.5.

6. State the factors affecting composting. State the different methods of composting. Explain any one in detail with neat labelled sketch. **(8M)**

Ans. Refer Sections 3.1, 3.1.1, 3.2 and 3.4.

7. State the factors affecting selection of site for landfilling of solid waste. State advantages and disadvantages of landfill method. **(8M)**

Ans. Refer Sections 3.3, 3.3.5 and 3.3.6.

8. Describe pyrolysis method of incineration w.r.t. meaning, principle, merits and demerits. **(8M)**

Ans. Refer Section 3.1.1.

9. Explain use of by-products of incineration method of solid waste. **(8M)**

Ans. Refer Section 3.10.5.

10. Explain the following: Leachate from landfilling. **(2M)**

Ans. Refer Section 3.3.8.

11. State methods of collecting recyclables. Describe current status of recycling of SW in India. **(8M)**

Ans. Refer Section 3.6 and 3.7.

Summer 2018

1. State the factors affecting composting process. **(4M)**

Ans. Refer Section 3.1.2.

2. Define recycling of solid waste and state the purposes of recycling. **(4M)**

Ans. Refer Sections 3.4.1 and 3.4.2.

3. State any two advantages and disadvantages of incineration of solid waste. **(4M)**

Ans. Refer Section 3.10.6.

4. State any four factors affecting the site selection for landfilling of solid waste. **(4M)**

Ans. Refer Section 3.3.6.

5. Enlist the methods of land filling of solid waste. Explain in brief about any one method. **(4M)**

Ans. Refer Section 3.3.7.

6. State the products of incineration process of solid waste and also state their uses. **(4M)**

Ans. Refer Section 3.10.5.

7. Define vermicomposting. Explain its concept in brief. **(4M)**

Ans. Refer Section 3.2.3.

8. Explain about Banglore method of manual composting of solid waste. **(4M)**

Ans. Refer Section 3.2.1.1.

9. Explain in brief present status of solid waste recycling in India. **(4M)**

Ans. Refer Section 3.7.

10. (i) Define Leachate. State the necessity to control it. **(4M)**
    (ii) Explain with neat sketch about single composite liner system and double liner system for the leachate control. **(4M)**

Ans. (i) Refer Sections 3.3.8 and 3.3.8.5.
(ii) Refer Section 3.3.8.5.

11. (i) State the various benefits of recycling of solid waste. **(4M)**
    (ii) Explain about various methods of collection of recyclable materials (solid waste). **(4M)**

Ans. (i) Refer Section 3.4.3.
(ii) Refer Section 3.6.

Winter 2018

1. State the typical constituents of municipal landfill gas. **(4M)**

Ans. Refer Table 3.2.

2. Define : (i) Pyrolysis, (ii) Leachate. **(4M)**

Ans. Refer Sections (i) 3.11 and (ii) 3.3.8.

3. Explain any one leachate control system with neat sketch. **(4M)**

Ans. Refer Section 3.3.8.5.

4. State the advantages and disadvantages of incineration process. **(4M)**

Ans. Refer Section 3.10.6.

5. State the types of incinerator and explain any one. **(4M)**

Ans. Refer Section 3.10.3.2.

6. Describe factor affecting composting process. **(4M)**

Ans. Refer Section 3.1.2.

7. State any four advantages and disadvantages of landfill method. **(4M)**

Ans. Refer Section 3.3.5.

8. Differentiate between Banglore method and Indore method of composting. **(4M)**

Ans. Refer Section 3.2.1.1.

9. Explain Vermicomposting. **(4M)**

Ans. Refer Section 3.2.3.

10. State methods of collection of recyclables and explain any one. **(4M)**

Ans. Refer Section 3.6.

11. State purpose of recycling of solid waste. **(4M)**

Ans. Refer Section 3.4.2.

Summer 2019

1. State any two advantages and disadvantages of incineration process. **(4M)**

Ans. Refer Section 3.10.6.

2. Describe Indore and Bangalore methods of composting. What is the main difference between them? **(4M)**

Ans. Refer Sections 3.1.1.1, 3.2.1.2 and 3.1.1.3.

3. How is the leachate management done? **(4M)**

Ans. Refer Chapter 3.3.8.7.

4. Define Pyrolysis and Gasification. **(4M)**

Ans. Refer Sections 3.11.1 and 3.12.

5. (i) Define landfill. **(4M)**
   (ii) State any four products of incineration process.

Ans. Refer Sections 3.3.1 and 3.10.5.

6. Describe the principles of composting. **(4M)**

Ans. Refer Section 3.1.1.

7. (i) What is vermi composting? **(4M)**
   (ii) State the various methods of pyrolysis.

Ans. Refer Sections 3.2.3 and 3.11.3.

8. How are the following recycled? **(4M)**
   (i) Fly ash, (ii) Blast furnace slags, (iii) Pulp and paper, (iv) Chromium Sludge

Ans. Refer Section 3.9.

9. Describe the area method of landfilling. Write its advantages and disadvantages. **(4M)**

Ans. Refer Sections 3.3.7.1 and 3.3.5.

10. How is the resources recovery through waste processing done? **(4M)**

Ans. Refer Section 3.9.

11. Enlist the common product that can be made with recycled contents. **(4M)**

Ans. Refer Section 3.5.

12. Explain sorting prior to waste processing or land filling. **(4M)**

Ans. Refer Section 3.3.2.

***

Chapter **4...**

Biomedical Waste Management and Health Aspects and Public Involvement in Solid Waste Management

Weightage of Marks = 14, Teaching Hours = 08

Syllabus

Biomedical Waste Management

4.1 Definition of Bio Medical Waste.

4.2 Sources and Generation of Biomedical Waste

4.3 Classification of Biomedical Waste

4.4 Management Technologies

Health Aspects and Public Involvement in Solid Waste Management

4.4 Health Aspects During Handling and Processing.

4.5 Health Problems During Time of Segregation, Recovery, Recycling and Reuse of Solid Waste.

4.6 Public Involvement and Participation in Solid Waste Management Practices.

About this Chapter

At the end of this chapter, students will be able to:

- Compare the characteristics of the given type of biomedical waste with other given type of waste material.
- Describe the different sources of generating the given type of biomedical waste in the given locality.
- Classify the biomedical waste on the basis of given criteria.
- Explain the method of executing the disposal of the given type of biomedical waste in the given area.
- Explain the ill effects on the health of humans handling the given type of solid waste in the given area.
- Justify the need of public participation in effective implementations of schemes managing the given type of solid waste in the given area.

4.0 OVERVIEW (W-16)

- Medical care is vital for our life, health and wellbeing. But the waste generated from medical activities can be hazardous, toxic and even lethal because of their high potential for diseases transmission.
- The hazardous and toxic parts of waste from health care establishments comprising infectious, bio-medical and radio-active material as well as sharps (hypodermic needles, knives, scalpels etc.) pose a grave risk, if these are not properly treated/disposed or are allowed to get mixed with other municipal waste.
- Its tendency to encourage growth of various pathogen and vectors and its ability to contaminate other non-hazardous/non-toxic municipal waste threatens the efforts undertaken for overall municipal waste management.
- The rag pickers and waste workers are often worst affected, because unknowingly or unwittingly, they dig through all kinds of poisonous material while trying to salvage items which they can sell for reuse.
- At the same time, this kind of illegal and unethical reuse can be extremely dangerous and even fatal. Diseases like cholera, plague, tuberculosis, hepatitis (especially HBV), AIDS (HIV), diphtheria etc. in either epidemic or even endemic form, pose grave public health risks.
- The rules framed by the Ministry of Environment and Forests (MoEF), Government of India, known as 'Biomedical Waste (Management and Handling) Rules, 1998,' notified on 20th July 1998, provides uniform guidelines and code of practice for the whole nation. It is clearly mentioned in this rule that the 'occupier' (a person who has control over the concerned institution/premises) of an institution generating bio-medical waste (e.g., hospital, nursing home, clinic, dispensary, veterinary institution, animal house, pathological laboratory, blood bank etc.) shall be responsible for taking necessary steps to ensure that such waste is handled without any adverse effect to human health and the environment.

4.1 BIOMEDICAL WASTE (S-16)

4.1.1 Definition of Biomedical Waste (S-15)

- "The waste generated by hospitals, nursing or maternity homes, clinics, dispensary, veterinary institutions, pathological laboratory, blood banks which is potentially infectious to human health and the environment is called as *Biomedical waste*".
- '*Biomedical waste*' means any solid and/or liquid waste including its container and any intermediate product, which is generated during the diagnosis, treatment or immunisation of human beings or animals or in research or in the production or testing.

Table 4.1: BMW Generation Rate (W-18)

Biomedical Waste	Non-hazardous (75-90%)		
	Hazardous (10-25%)	Infectious (15-18%) • Non-sharp • Sharp • Liquid waste	Other hazardous (05-07%) • Radioactive waste • Discarded glass • Chemical waste • Cytotoxic waste

4.1.2 Components of Biomedical Waste (S-15, 19)

- Human anatomical waste (tissues, organs, body parts etc.).
- Animal waste (as above, generated during research/experimentation, from veterinary hospitals etc.)

- Microbiology and biotechnology waste, such as, laboratory cultures, micro-organisms, human and animal cell cultures, toxins etc.
- Waste sharps, such as, hypodermic needles, syringes, scalpels, broken glass etc.
- Discarded medicines and cyto-toxic drugs.
- Infectious waste like soiled waste, such as dressing and bandages, plaster casts, pathological tissues, material contaminated with blood etc.
- Solid waste or plastic waste (disposable items like IV sets, tubes, catheters, blood bags, syringes etc. excluding sharps).
- Sharps means needles, broken glass, blades, razors etc.
- Liquid waste generated from any of the infected areas.
- Incineration ash.
- Chemical waste.
- Cytotoxic waste
- Radioactive waste etc.

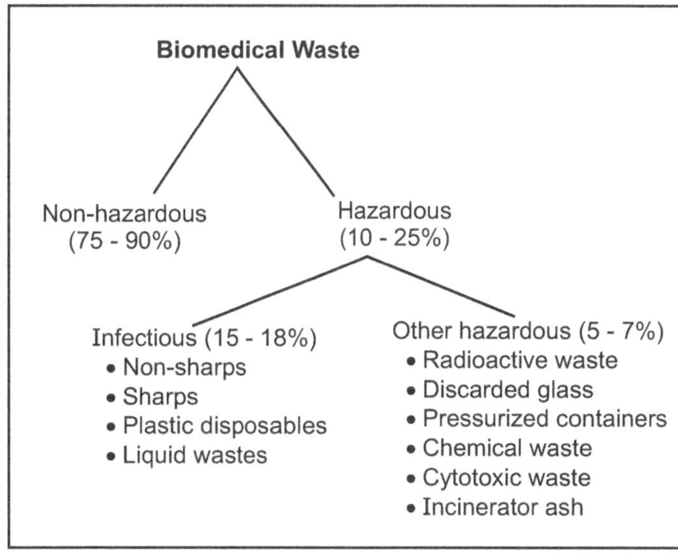

Fig. 4.1: Classification of Biomedical Waste

4.2 SOURCES OF GENERATION OF BIOMEDICAL WASTE (S-15, 16; W-16, 18)

- Biomedical waste is generated from biological and medical sources and activities, such as the diagnosis, prevention, or treatment of diseases and medical research centres. Common generators (or sources) of biomedical waste include:

1. Hospitals,
2. Health Clinics,
3. Nursing Homes,
4. Medical Research laboratories,
5. Veterinary homes.
6. Pharmacies,
7. Blood banks,
8. Medical research laboratories.
9. Funeral homes,
10. Offices of physicians, dentists and veterinarians.

4.3 CLASSIFICATION OF BIOMEDICAL WASTE (S-15, 16, 18, 19)

Table 4.2

Option	Waste Category	Treatment and Disposal
Category No. 1	Human Anatomical Waste (Human tissues, organs, body parts)	Incineration@/deep burial*
Category No. 2	Animal Waste (Animal tissues, organs, body parts carcasses, bleeding parts, fluid, blood and experimental animals used in research, waste generated by veterinary hospitals, colleges, discharge from hospitals, animal houses)	Incineration@/deep burial*
Category No. 3	Microbiology and Biotechnology Waste (Wastes from laboratory cultures, stocks or specimens of micro-organisms live or attenuated vaccines, human and animal cell culture used in	Local autoclaving/micro-waving/incineration@

Option	Waste Category	Treatment and Disposal
	research and infectious agents from research and industrial laboratories, wastes from production of biologicals, toxins, dishes and devices used for transfer of cultures)	
Category No. 4	Waste sharps (Needles, syringes, scalpels, blades, glass etc. that may cause puncture and cuts. This includes both used and unused sharps)	Disinfection (chemical treatment@01/auto claving/micro-waving and mutilation/ shredding)
Category No. 5	Discarded Medicines and Cytotoxic drugs (Wastes comprising of outdated, contaminated and discarded incineration@/destruct ion and medicines)	Incineration@/destruct ion and drugs disposal in secured landfills.
Category No. 6	Solid Waste (Items contaminated with blood, and body fluids including cotton, dressings, soiled plaster casts, lines, beddings, other material contaminated with blood)	Incineration@ autoclaving/microwaving
Category No. 7	Solid Waste (Wastes generated from disposable items other than the waste sharps such as tubings, catheters, intravenous sets etc).	Disinfection by chemical treatment @@autoclaving/ microwaving and mutilation/ shredding##.
Category No. 8	Liquid Waste (Waste generated from laboratory and washing, cleaning, house-keeping and disinfecting activities)	Disinfection by chemical treatment @@and discharge into drains.
Category No. 9	Incineration Ash (Ash from incineration of any bio-medical waste)	Disposal in municipal landfill.
Category No. 10	Chemical Waste (Chemicals used in production of biologicals, chemicals used in disinfection, as insecticides etc.)	Chemical treatment@@ and discharge into drains for liquids and secured landfill for solids.

@@ Chemical treatment using at least 1% hypochlorite solution or any other equivalent chemical reagent. It must be ensured that chemical treatment ensures disinfection.

Mutilation/shredding must be such, so as to prevent unauthorised reuse.

@ There will be no chemical pretreatment before incineration. Chlorinated plastics shall not be incinerated.

* Deep burial shall be an option available only in towns with population less than five lakhs and in rural areas.

4.4 PROVISIONS IN THE LAW (S-16, 18; W-15, 16)

- Safe disposal of biomedical waste is now a legal requirement in India. The ministry of Environment and Forests notified the Biomedical Waste (Management and Handling) Rules, 1998 in July 1998.
- In accordance with these rules, following provisions are made:
  - It is the duty of every "occupier" i.e. a person who has the control over the institution or its premises, to take all steps to ensure that waste generated is handled without any adverse effect to human health and environment.

- The hospitals, nursing homes, clinics, dispensaries, pathological laboratories etc., are therefore required to set in place the biomedical waste disposal and treatment facilities as per Schedule V and VI respectively. It is however not incumbent that every institution has to have its own waste treatment facility.

- The rule also envisages that common facility or any other facilities can be used for waste treatment. However it is incumbent on the occupier to ensure that the waste is treated within a period of 48 hours. Biomedical Waste (Management and Handling) Rules have six schedules as briefed in Table 4.3.

Table 4.3: Biomedical Waste Rules (S-15, 18; W-18)

Schedule	Contents
Schedule I	Classification of biological waste in various categories.
Schedule II	Colour coding and types of containers to be used for each category of biomedical waste.
Schedule III	Proforma of the label to be used on container/bag.
Schedule IV	Proforma of the label for transport of waste container/bag.
Schedule V	Standards for treatment and disposal of wastes.
ScheduleVI	Deadline for creation of waste treatment facilities.

4.5 OBJECTIVES OF BIOMEDICAL WASTE MANAGEMENT (W-15)

1. To avoid transmission of disease from patient to patient, from patient to health worker and vice versa.
2. To avoid injury to the health care worker and workers in support services, while handling biomedical waste.
3. To stop general exposure to the harmful effects of the cytotoxic, genotoxic and chemical biomedical waste.
4. Define medical waste, regulated and medical waste and infectious waste.
5. Discuss the regulations applicable to medical waste.
6. Discuss the components of an infectious waste management plan.
7. Outline an exposure control plan.
8. Discuss steps to take if exposed to infectious waste.
9. Discuss the problem of mercury.
10. Discuss records to maintain.
11. Recommend inspection items.
12. Discuss use of contractors.

Biomedical Waste Management Techniques: (W-18; S-19)

Biomedical waste management technologies are as follows.

1. Deep Burial
2. Autoclaving
3. Microwaving
4. Chemical disinfection
5. Mutilation/Shredding
6. Incineration
7. Secured Sanitary landfills
8. Encapsulation

Deep Burrial: It is a pit 2 m deep half filled with waste and remaining with lime within 50 cm of the surface before filling the rest with soil. Every time when biomedical wastes are added to the pit a layer of 10 cm of soil should be added to cover the waste. The deep burial site should be relatively impermeable, away from residential area, vicinity of drinking water to avoid risk of pollution. The institute should maintain the record of all sites of deep burial.

Autoclaving: It involves sterilization with steam at 120°C temperature under 15 psi pressure for 60 to 90 minutes. Autoclaves are used for disinfection of surgical instruments. Infectious waste can be autoclaved by exposing it to 160°C temperature at 6 bars pressure for about 20 minutes.

4.6 BIOMEDICAL WASTE MANAGEMENT TECHNOLOGIES/PROCESSES (W-18)

- Handling, segregation, mutilation, disinfection, storage, transportation, final disposal and public education are vital steps for safe and scientific management of BMW in any establishment.

Methods of Collection and Segregation of Biomedical Waste: (S-15)

Following are the various methods of collection and segregation of BWM in colour coded bags or containers:

1. **Yellow bags:** All the infectious wastes are collected into these bags. It may include soiled bandage, cotton or any other thing which is infected. The waste from these bags is treated by incineration only.
2. **Red bags:** All the plastic wastes such as injections, syringes, I. V. tubing, Bottles are collected into these bags. They are treated by incineration only.
3. **Blue bags:** It consists of all the types of glass bottles and broken glass articles. It is classified as hazardous waste and treated by incineration only.
4. **Black carboy:** All sharps of metal are collected into these bags like blades, needles without syringes etc.
5. **Purple container:** It consists of cytotoxic and cytostatic wastes, clinical waste. It is hazardous waste and has to be treated by incineration only.
6. **Yellow and Black containers:** It consists of offensive but hygiene waste from health centers. It is non-hazardous and can be treated by incineration, landfill.
7. **Black containers:** It consists of domestic waste mixed with municipal waste which is classified as non-hazardous waste.
8. **White containers or bags:** It consists of dental amalgam and mercury such as spent and expired capsules.

(a) Waste Collection and Segregation

- Waste should be collected and segregated at the site generation itself. Its aim is to keep the harmful waste separate from the harmless and non-contagious waste. The key to minimization and effective management of biomedical waste is segregation and identification of the waste. The most appropriate way of identifying the categories of biomedical waste is by sorting the waste into colour coded plastic bags or containers in accordance with schedule II of Biomedical Waste (Management and Handling) Rules.

Table 4.4 (W-15)

Colour Coding	Type of Container	Waste Category	Treatment Option
Yellow	Plastic bag	Cat. 1,2,3,6	Incineration/deep burial.
Red	Disinfected container/plastic bag	Cat. 3,6,7	Autoclave/microwave/chemical treatment.
Blue/white translucent	Plastic bag/puncture proof	Cat. 4,7	Autoclave/microwave/chemical treatment and destruction shredding.
Black	Plastic bag	Cat. 5,9,10	Disposal in secure landfill.

Notes:

- Colour coding of waste management with multiple treatment options shall be selected depending on the treatment option chosen.
- Waste collection bags for waste types needing incineration shall not be made of chlorinated plastics.
- Categories 8 and 10 (liquid) do not require container/bags.
- Category 3 if disinfected locally need not be put in containers/bags.

Procedure for Waste Collection:

- Specifically coloured plastic bag should be kept in its container. Bins and bags should bear the biohazard symbol.
- As soon as three fourth of the bag is full of waste it should be removed from the container, tied tight with a plastic string and properly labelled.
- Under no circumstances, an infectious waste should be mixed with the non-infectious waste.
- Collection of disposable items (syringes, I/V bottles, catheters, rubber gloves etc.) should be undertaken when they have been mutilated (cut) chemically disinfected (by dipping in 1% hypochlorite solution for 30 min.)
- Syringe barrel should always be separated from the plunger before disinfection. Needles should be destroyed with needle destroyer. Manual mutilation of sharps should never be tried as it may cause injury. All other sharps must be strongly disinfected (chemically) before they are shredded or finally disposed. Sharps should be kept in puncture proof containers and properly labelled.
- Biomedical waste handlers should be trained in handling the waste and made aware of proper way of handling waste to avoid injury and accidents.

(b) Transportation and Storage

- The waste may be temporarily stored at the central storage area of the hospital and from there it may be sent in bulk to the site of final disposal once or twice a day depending upon the quantum of waste.
- During transportation, following points should be taken care of:
    - Ensure that waste bags/containers are properly sealed and labelled.
    - Bags are picked up from the neck and placed so that bags can be picked up by the neck again for further handling. Hand should not be put under the bag. At a time only one bag should be lifted.
    - Manual handling of waste bags should be minimized to reduce the risk of needle prick injury and infection.
    - Biomedical waste should be kept only in a specified storage area.
    - After removal of the bag, clean the container including the lid with an appropriate disinfectant.
    - Waste bags and containers should be removed daily from wards/OPDs or even more frequently if needed (as in Operation Theatres, ICUs, labour rooms). Waste bags/containers should be transported in a covered wheeled containers or large bins in covered trolleys.
    - Biomedical waste storage area should be separate from the general waste storage area.

(c) Central Storage

- The central storage area in a hospital should be ideally situated on the ground floor near the rear entrance. This makes the transportation of waste to the site of final disposal easier.
- The central storage area should have sufficient storage capacity to store the required number of waste bags, depending upon the quantum of waste generated in the hospital. It should have the storage capacity of at least 2 days' waste.
- It should have good flooring, light, ventilation, water supply and drainage system.
- A full time storekeeper should be there to receive and dispatch the waste and to maintain proper record.
- As per rules biomedical waste can not be stored for more than 24 to 48 hrs. (Refrigerated storage room should be available where wastes have to be stored in bulk for over 48 hrs.)

(d) Transport to Final Disposal Site

- Transportation from health care establishment to the site of final disposal in a closed motor vehicle (truck, tractor-trolley etc.) is desirable as it prevents spillage of waste on the way. Vehicles used for transport of biomedical waste must have the "Bio-Hazard" symbol and these vehicles should not be used for any other purpose.

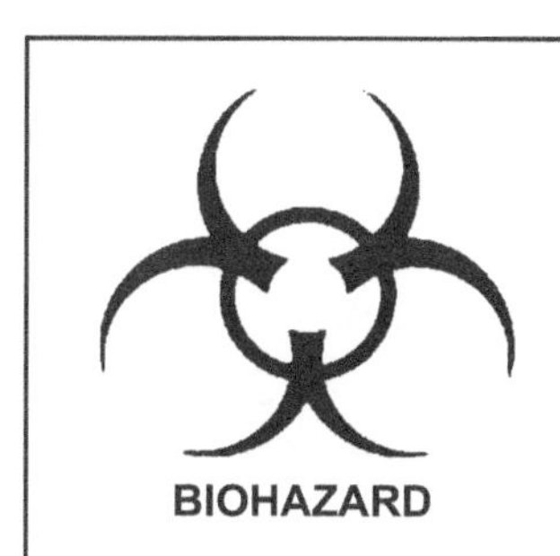

Fig. 4.2

(e) Disposal of Biomedical Waste Pretreatment

- The infected waste that can not be incinerated (e.g. Plastic and rubber items, sharps) has to be disinfected first, before it is sent for final disposal.

(f) Final Disposal

- Incineration is a high temperature dry oxidation process, which reduces organic and combustible waste to inorganic incombustible matter. This method is usually used for the waste that can not be reused, recycled or disposed off in landfill site.

Characteristics of Waste Suitable for Incineration:

- Low heating volume – above 2000 kCal/kg for single chamber incinerators and above 3500 kCal/kg for pyrolytic double chamber incinerators.
- Content of combustible matter above 60%.
- Content of non-combustible matter below 50%.
- Content of non-combustible fines below 20%.
- Moisture content below 30%.

Waste Types not to be Incinerated:

Waste types not to be incinerated are:

- Pressurized gas containers.
- Large amount of reactive chemical wastes.
- Silver salts and photographic or radiographic wastes.
- Halogenated plastics such as PVC.
- Waste with high mercury or cadmium content such as broken thermometers, used batteries.
- Sealed ampoules or ampoules containing heavy metals.

(g) Types of Incinerators

- **Single chamber furnaces with static grate:** These should be used only if pyrolytic incinerators are not affordable.
- **Double Chamber Pyrolytic Incinerators:** These are the most commonly used incinerators. In the first (pyrolytic) chamber, waste is destroyed through an oxygen deficient, medium temperature combustion process (800°C).This produces solid ashes and gases. In the second chamber, gases are burnt at a high temperature (900-1200°C) using an excess of air to minimize smoke and odour. This type of incinerator is somewhat expensive and requires trained personnel to handle it.
- **Rotary Kilns:** It comprises of rotating oven and a post combustion chamber. They are used to burn chemical wastes (chemicals, pharmaceuticals including cytotoxic drugs).

(h) Safe Pit for Sharps

- Sharps (needles and blades etc.) are being used in a day-to-day practice in all health care establishments. To avoid recycling of sharps, their burial in safe pit is an effective and economical disposal method. It can be constructed by 5 feet deep circular concrete ring of 3′ diameter. A slab is used on top in which GIC pipe with 5″ or 6″ diameter is used which is fitted by lock and key arrangements. Size of the pit may vary as per the quantum of sharp waste to be disposed off. The pit is plastered inside at bottom and around. When it is filled up, cement slurry can be used to close it and second pit is constructed.

(i) Landfill Disposal

- It is another method of final disposal of biomedical waste. If a municipality or medical authority genuinely lacks the means to treat the waste before disposal, sanitary landfill observing certain standards can be as an acceptable choice especially in developing countries.

Standards for Deep Burial:

- A pit or trench should be dug about 2 metres deep. It should be half filled with waste, then covered with lime within 50 cm of the surface, before filling the rest with soil.
- It must be ensured that animals do not have any access to burial site.

- On each occasion, when wastes are added to the pit, a layer of 10 cm of soil shall be added to cover the wastes.
- Deep burial site should be relatively impermeable and no shallow well should be close to the site. The site should be away from the residential area and the vicinity of drinking water so as to avoid the risk of pollution.
- The location of deep burial site will be authorized by the prescribed authority.
- The institution shall maintain a record of all pits for deep burial.

(j) Awareness and Education

- In India, hospitals and other health care establishments are not well equipped to handle the enormous amount of biomedical waste. There is an urgent need to raise the awareness amongst all concerned. Information can be circulated through organizing seminars, workshops, practical demonstrations, group discussions, lectures etc.
- It is vital to formulate an effective education and training programs specific for different target groups involved in biomedical waste handling and management.

Disposal of Biomedical Waste: (S-17)

1. **Incineration:** Incineration is typically used for pathological and pharmaceutical waste. Incineration of medical waste should be performed in a controlled facility to ensure complete combustion and minimize any negative effects for the environment.
2. **Land Disposal:** Land disposal is typically used for shredded, treated and decontaminated waste. In certain cases, it can also be used for hazardous waste or other untreated waste that can not be decontaminated by other means. Specialized sanitary landfill sites exist to reduce the risk of soil and water contamination and provide a safe space for medical waste disposal.
3. **Deep Burial:** A pit or trench is excavated about 2 m deep. It is half filled with waste, and then covered with lime within 50 cm of the surface, then filling the remaining portion with soil. Cover is provided at top with lock facility.
4. **Autoclaving:** Thermal treatment is typically used for sharp and certain other types of infectious waste. An autoclave is like a large pressure cooker that uses high temperatures and steam to deeply penetrate all materials and kill any microorganisms. Depending on the type and amount of waste one should need to sterilize. Modern autoclaves are also automated to minimize human involvement and therefore reduce needle-stick injuries and contamination.
5. **Chemical Treatment:** Often used to deactivate liquid waste, chemical treatment is designed to decontaminate or deactivate certain wastes on site rather than packaging and sending them to a separate facility.

   Chemical treatment can also be applied to some non-liquid infectious wastes, but they would typically need to be shredded first to ensure that all portions of the waste are exposed to the chemicals.

   Depending on the type of waste, chemicals like chlorine, sodium hydroxide or calcium oxide can be used. Chemical treatment has to be executed carefully and by knowledgeable staff.
6. **Microwave Treatment:** A microwave treatment system, similar to an autoclave, also uses heat to decontaminate medical waste. These systems work best for waste that is not 100% dry or solid, as the moisture allows the heat to penetrate deeper, and the steam sterilizes.

Effects of Biomedical Waste: (S-18)

Wildlife and Pharmaceuticals:

- Biomedical waste that is not disposed off properly can end up in lakes, parks and other wildlife refuges where birds and fauna live.
- Wildlife is very curious about pharmaceuticals. It is thought, they are attracted to the scent or colour of pills and liquid medicine. This curiosity results in digestion of medication, which can injure or even kill the animal.

Groundwater Contamination:

- Much thought and effort has been taken to ensure landfills are built to protect the earth around them. Most are built with a special lining so nearby soil and groundwater cannot become contaminated.
- Mishandled biohazard waste can compromise even the best landfill design. Syringes and other sharp objects can easily rip the lining. As rain falls, contaminants in the landfill can seep out to the exterior soil, and the groundwater become toxic.

Radioactive Pollution:

- In order to accurately diagnose patients, doctors must sometimes use radioactive tools. When disposed off improperly, radioactivity can enter landfills and other areas. These substances emit particles that are dangerous to people. Excessive exposure to radioactivity can result in serious diseases.

Airborne Pollutants:

- Certain medical waste can be destroyed by incineration. But, if not ignited properly, pollutants can move through the air.
- Airborne pollutants can be worse than land-based types because they can spread far and wide and quickly.

4.7 HEALTH ASPECT DURING HANDLING AND PROCESSING (S-18; W-15, 16, 18)

- There are potential risks to environment and health from improper handling of solid wastes. Direct health risks concern mainly the workers in this field, who need to be protected, as far as possible, from contact with wastes.
- There are also specific risks in handling wastes from hospitals and clinics.
- For the general public, the main risks to health are indirect and arise from the breeding of disease vectors, primarily flies and rats.
- The most obvious environmental damage caused by municipal solid wastes is aesthetic, the ugliness of street litter and degradation of the urban environment and beauty of the city.
- More serious, however, and often unrecognised, is the transfer of pollution to water, ground water.
- Air pollution can be caused from the inefficient burning of wastes, either in open air, or in plants that lack effective treatment facilities from the gaseous effluents.
- Uncontrolled hazardous wastes from industries mixing up with municipal wastes create potential risks to human health.
- Traffic accidents can result from toxic spilled wastes.
- There is specific danger of concentration of heavy metals in the food chain, a problem that illustrates the relationship between municipal solid wastes and liquid industrial effluents containing heavy metals discharged to a drainage/sewerage system and/or open dumping sites of municipal solid wastes and the wastes discharged thereby maintains a vicious cycle.
- Municipal Solid Waste Management System involves various activities like storage, collection, transportation, disposal etc. These activities even if properly controlled and with proper precautionary measures adopted, may have adverse impact on land, water and air environment, human and environmental health, aesthetics and quality of life.
- Epidemiological studies have shown that a high percentage of workers who handle refuse, and of individuals who live near or on disposal sites, are infected with gastrointestinal parasites, worms and related organisms.
- Contamination of this kind is likely at all points where waste is handled.
- Although it is known that, vector insects and rodents can transmit various pathogenic agents (amoebic and bacillary dysentery, typhoid fever, salmonellosis, various parasites, cholera, yellow fever, plague and others), it is often difficult to trace the effects of such transmission to a specific population.

- During the last decade of the 19th century as well as during the 5 initial years of 20th century, millions of people died due to Bubonic Plague in India, which had linkages to poor management of Solid Waste.
- More recently a study by the US Public Health Service has demonstrated the relationship of 22 human diseases to improper solid waste management.
- The organic fraction of Municipal Solid Waste is an important component, not only because it constitutes a sizable fraction of the solid waste stream, but also because of its potentially adverse impact upon public health and environmental quality.
- A major adverse impact is due to its attraction of rodents and vector insects for which it provides food and shelter. Impact on environmental quality takes the form of foul odours, unsightliness, land, water, air and noise pollution.
- These impacts are not confined merely to the disposal site. On the contrary, they permeate the area surrounding the site and wherever the wastes are generated, spread or accumulated.
- Unless an organic waste is appropriately managed, its adverse impact will continue until it has fully decomposed or otherwise stabilized. Uncontrolled or poorly managed intermediate decomposition products can contaminate air, water and soil resources.

Types of Environmental and Health Hazards: (W-15)

- Chemical
- Physical
- Biological
- Psychological

4.7.1 Impact of Solid Waste on Human Health and Environment (S-15)

Impacts of Solid Waste on Health:

- Low birth weight.
- Cancer.
- Neurological disease.
- Increase in mercury level in fish due to disposal of mercury in the rivers. This is harmful for human health.
- Chemical poisoning through chemical inhalation.
- Nausea and vomiting.
- Increase in hospitalization of diabetic resident living near waste site.

Impacts of Solid Waste on Environment:

- Waste breaks down in landfills to form methane, which causes greenhouse gas.
- Change in climate and destruction of ozone layer due to biodegradable waste, due to waste pollutions and illegal dumping.
- Leaching is a process by which solid waste enter soil and ground water, and contaminate them.

4.8 HEALTH PROBLEMS DURING TIME OF SEGREGATION, REUSE, RECOVERY, RECYCLING OF SOLID WASTE (S-16; W-16)

- Infrequent collection and rapid decomposition of wastes provide an attractive feeding and breeding site for flies, rats and other scavengers. Human and animal faecal matter or hospital wastes are often mixed with the refuse. Domestic and on occasion industrial, solid wastes are disposed off in open spaces within residential areas.
- Refuse disposal is often a non-profit making business and thus is treated as an unwanted side-effect of development. Attention should be paid to safe storage, collection, transport and intermediate transfer to bulk transport, and final disposal.
- In many places waste recovery is an important unorganized private industry employing many thousands of scavengers who may live or work on refuse dumps.
- They are referred to as human scavengers or waste pickers and are frequently ignored in urban project plans although their activities may be vital to the life of the city. Many consist of abandoned children and destitute families.

- They live and work under extensive health risks, which are largely undocumented, and suffer severe exploitation and deprivation. Possible health hazards include raised levels of infant mortality, hand and leg injuries, intestinal and respiratory infections, eye infections, lower back pain, malnutrition, skin disorders and exposure to hazardous waste.
- Water supply, for drinking and washing, and sanitation facilities are usually very poor at dumpsites.

Health Problems during Segregation and Recycling Of Solid Waste: **(W-16, 18)**

- During segregation and recycling of solid waste sanitary workers face to direct health risks.
  1. Workers and rag pickers can be infected while picking of waste and may have communicable diseases such as diarrhoea and dysentery (due to flies), skin diseases etc.
  2. If biodegradable i.e. wet waste and non-biodegradables i.e. dry waste are not put separately, it can create bad odour.
  3. Possible health hazard include raised level of infant mortality, non-communicable disease such as hand/leg injury by sharp edge material, respirational infections, eye infection etc.
  4. For public, health problems arises from the breeding of diseases due to flies, insects and rats. The health problems are more serious when transfer of pollution to water, ground water and air takes place.
  5. Air pollution causes due to insufficient burning of wastes in open or in plants due to improper gaseous effluent removal devices.
  6. Mixing of hazardous wastes from industries with municipal wastes create risks to human health. Danger of concentration of heavy metals in the food chain due to industrial effluent discharges into drainage system creates danger to public health.
  7. Sanitary workers dealing with municipal solid waste are infected with gastrointestinal parasites, worms etc. The organic fraction of municipal solid waste provides food and shelter to insects and rodents which causes 22 human diseases due to improper solid waste management causing adverse impact upon public health and environmental quality.

4.8.1 Various Health and Environmental Problems due to Solid Waste **(S-15)**

4.8.1.1 Communicable Diseases

- *Communicable diseases* are the diseases that spread from one person to another through airborne bacteria or viruses or through direct contact with infected person.
- Houseflies may be important in the transmission of enteric infections, particularly those responsible for infantile diarrhoea and dysentery.
- Disease transmission by houseflies is greatest where inadequate refuse storage, collection and disposal (leading to increased breeding) is accompanied by inadequate sanitation.
- Thus, flies gain greater access to human faeces and then to food. Refuse must be collected daily to prevent fly breeding.

4.8.1.2 Non-Communicable Disease **(S-15)**

- *Non-communicable diseases* are the diseases that are cannot transmitted from infected person to other person through air, water or direct contact and hence are non-infectious.
- Once collected in poorly operated disposal sites, rubbish may contaminate groundwater with nitrates, heavy metals and other chemicals.
- Incineration of wastes may pollute the air with particulates and oxides of sulphur and nitrogen.
- The slag and ashes from incinerators may result in leachates that are rich in heavy metals and other potentially toxic substances.

4.8.1.3 Injury

- Combustible gases will be generated from waste landfills for more than 20 years and these travel under roads and through ducts to create a hazard in buildings and land fill sites.
- People collecting rubbish may be injured by sharp objects including glass, metal and wood. These may lead to puncture wounds and gashes which may become infected and cause serious morbidity.
- Composted solid waste can cause injury to farmers as sharp objects are not always properly removed.

4.5.1.4 Aesthetic Aspect

- Foul odour is emitted at the disposal site due to continuous decomposition of organic matter and emission of methane, hydrogen sulphide, ammonia etc. The problem is intensified if proper mitigation measures are not adopted.
- Odour is also emitted at the collection points if quick removal of wastes is not practised. Spreading of the waste in the area adjacent to the dustbin due to activity of ragpickers cause degradation of aesthetic quality. Uncontrolled disposal and open burning of wastes at the landfill sites create poor vision.
- Domestic rats, birds and other scavenging animals act as reservoirs for many organisms transmissible to people, including plague, forms of typhus, leptospirosis, trichinosis, psittacosis, salmonella infection and bovine tuberculosis.
- Chemical control of both houseflies and rodents is not very effective because of widespread resistance. The essential basis of control remains denial of access to food and harborage, by covered storage and efficient removal.
- Aedes mosquitoes, vectors of dengue and yellow fever, breed prolifically in discarded containers that trap rainwater. Culex mosquitoes, vectors of filariasis, breed in polluted stagnant water. Such breeding sites often occur where drains are choked by solid waste.

Health Safety Measures During Handling and Processing of Solid Waste: **(S-17)**

1. Workers on field should wear hand gloves.
2. They should wear safety shoes.
3. To avoid polluted air they should wear mask.
4. Every month complete medical checkup should be done.
5. They should wear safety goggles to avoid eye disease.
6. Separation of solid waste should be done according to category (dry, wet, medical, commercial, residential etc.).

Table 4.3: Types of Environmental and Health Hazards

Environmental and Health Hazards	Examples and Causes
1. Environmental pollution	1. Air quality, water quality, land use, noise
2. Communicable disease	2. Gastrointestinal disorders, diarrhoea, respiratory infection, skin diseases, jaundice, trachoma, eosinophilia etc.
3. Non-communicable disease	3. Poisoning, hearing defects/loss, dust.
4. Injury	4. Occupational injury by sharps, needles, glasses, metals, wood, violence etc.
5. Aesthetics	5. Odour, visibility, dust etc.

4.9 PUBLIC INVOLVEMENT AND PARTICIPATION IN SOLID WASTE MANAGEMENT (S-16, 19; W-18)

- Public is in the centre of all the activities, yet it is ignored by the decision makers and made to merely wait and watch and ultimately what people get in hand is what they do not want or what is not in their priority. This creates a gap between the administrators and those administered and an atmosphere of apathy is created which distances people from government initiatives.
- Public awareness, effective community participation, transparent and clean administration, introduction of citizen charters and accountability at all levels can only bridge this gap.
- Solid Waste Management (SWM) is one such activity, where public participation is key to success. The local body can never be successful in Solid Waste Management without active community participation, whatever may be the investments made from the municipal or Government funds.

- The local bodies are the institutions of grass root democracy having elected members representing as small group of electorate. It also has an outreach service at the ward level through which it can easily interact with the people on almost all-important issues. The local body should therefore, seriously consider involving community in all programmes through a consultative process and variety of other communication approaches dealt with in this chapter later and adopt the strategy which has the acceptance of the community.

Importance of Public Involvement and Participation in SWM: (S-15)

Public involvement and participation in SWM is very important because of following points:

1. To increase the awareness of solid waste management among the people.
2. To increase the efficiency and effectiveness of planning process and implementation of solid waste management.
3. To understand the planning importance and significance.
4. To play an important role in the permitting process in case of hazardous waste as well as municipal waste facilities.
5. To improve the waste management strategies, negotiations with municipal authorities for better involvement in decision making.
6. To achieve the 3R principles.
7. To reduce littering of waste on streets and into drains, open spaces etc.
8. To encourage and assist the local composting and recycling initiatives.

4.10 STRATEGY OF PUBLIC PARTICIPATION (W-15)

- The following strategy may be adopted by the Urban Local Bodies (ULBs).

4.10.1 Identification of People's Groups

4.10.1.1 Residential Areas

Community may be classified into three categories:

1. High Income Group: The affording
2. Middle Income Group: Educated, sensitive, less affording
3. Low Income Group: Un-affording

- Sensitivity of each group is different and therefore needs to be tackled differently.
- Public involvement can be increased by public awareness programs, effective community participation, transparent and clean administration, introduction of citizen chapters and accountability at all levels.

Steps of Public Involvement and Participation in SWM: (W-18)

Public participation in solid waste management can be achieved by following steps.

1. Identification of areas where public participation is essential.
2. Reaching the community and seeking the co-operation for problem finding and their solution.
3. Public information, education and communication programs by group meetings, workshops, exhibitions, lectures and panel discussions.
4. Use of print media, TV, Cinema, Radio, Web sites, Posters, Pamphlets, School curriculum, NGO etc.

Importance of Public Involvement and Participation:

1. To increase the awareness of SWM among the people.
2. To promote the principles of reduction, reuse, recycling and recovery.

3. To reduce littering of waste on streets, drains and open spaces.
4. To plan segregation of hazardous, non-hazardous; dry wet waste, E-waste, BMW and industrial waste.
5. To reduce the environmental pollution.
6. To increase efficiency of planning and implementation of SWM.
7. To improve management strategies with local municipal authorities.
8. To encourage composting and recycling initiatives in public and agencies.

4.10.1.2 Markets/Commercial Areas/Offices/Banks etc.

These places may be classified into three broad categories:

1. Vegetable Markets
2. Shopping areas
3. Offices/Institutional areas

Different Methods of Communication for Mass Education: (S-19)

Following are the methods of communication for mass education

(i) Use of print media
(ii) Use of TV/Cable TV/ Radio
(iii) Street plays, puppet show
(iv) Posters
(v) Pamphlets
(vi) Use of hoarding
(vii) Use of public transport system
(viii) Web site/smart phones

4.10.2 Identification of the Areas in Solid Waste Management where Public Participation is Essential

4.10.2.1 People's Participation is Essential in the Following Areas

1. Reduce, Reuse and Recycling (R-R-R) of waste.
2. Not to throw the waste/litter on the streets, drains, open spaces, waterbodies etc.
3. Storage of organic/bio-degradable and recyclable waste separately at source.
4. Primary collection of waste.
5. Community storage/collection of waste in flats, multi-storeyed buildings, societies, commercial complexes etc.
6. Managing excreta of pet dogs and cats appropriately.
7. Waste processing/disposal at a community level (optional).
8. Pay adequately for the services provided.

The Importance of Public Participation in Solid Waste Management: (S-19)

Public participation in SWM is very important because of following points:

1. To increase the awareness of solid waste management among the people.
2. To increase the efficiency and effectiveness of planning process and implementation of solid waste management.
3. To understand the planning importance and significance.
4. To play an important role in the permitting process in case of hazardous waste as well as municipal waste facilities.
5. To improve the waste management strategies, negotiations with municipal authorities for better involvement in decision making.
6. To achieve the 3R principles.
7. To reduce littering of waste on streets and into drains, open spaces etc.
8. To encourage and assist the local composting and recycling initiatives.

4.10.3 Reach the Community

- The local body should decide the methodology to be adopted for reaching the community and seeking their co-operation and effective participation in SWM services. This is a very difficult area of activity and unless this is done meticulously, desired results will not be achieved.
- The essential step in this direction is to select representative samples of the community and go through a consultative process to ascertain the perceptions of the people about the SWM services being given to them, their expectations and extent to which they are willing to support and participate in the process. Their choice of technological options available also needs to be ascertained.
- The consultative process could be taken up as under:

4.10.3.1 Identification of Problems

- Identification of problems of waste management through site visits and consultation with local population at the time when the community is generally available for interaction. It may either be in early morning or late evening. The areas may be selected by following the method of drawl of representative samples.
- Situation analysis may be done by the persons who know the subject reasonably well, know local language and can communicate with local population effectively. Such persons may be Non-Governmental Organisations, Community Based Organisations or knowledgeable individuals. They should try to find out the prevalent situation of waste management in the area under observation and ascertain the perceptions of the people about the services provided. In this exercise, the local councillors, local leaders, NGOs etc., may be invited to participate.

4.10.3.2 Finding out Optional Solutions

- Having identified the deficiencies in the system and known the public perceptions, the next essential step is to think of optional solutions to tackle the problems, workout the cost implications and level of public participation needed.

4.10.3.3 Consult Community on Options Available

- Having done this homework, there should be second round of consultative process where the options worked out may be discussed with the community alongwith cost implications and their support required. Their suggestions may be sought on each solution proposed. The community may be encouraged to give their views freely.
- If we ask the people straight away the solutions of the problems, they may not be able to give right kind of suggestions as they have no exposure of various technological options. They must, therefore, be first appraised of the options available and then asked to give their considered opinion on 'what will work in their area' and 'how much they are willing to co-operate'.

4.10.3.4 Workout the Strategy of Implementation

- After the consultative process, strategy for implementation of the system may be worked out and pilot projects may be taken up in the areas where better enthusiasm is noticed and demonstrate the successes to other areas and gradually implement in rest of the areas of the city/town. It is desirable to implement the new program in a few areas to begin with, monitor its success carefully and extend the program to other areas thereafter with suitable modifications wherever necessary.

4.10.4 Public Information, Education, Communication Programs (IEC)

- For the successful implementation of any program involving public at large in SWM system, it is essential to spell out clearly and make them known the manner in which local body proposes to tackle the problem of waste management and extent to which public participation in Solid Waste Management is expected to keep the city clean and improve the quality of life in the city.

- Dayal Committee of Government of India (Report 1995) has advocated that IEC approach should:
  - Ensure that the people become aware of the problems of waste accumulation and the way it affects their lives directly.
  - Ensure that the people generate less waste by cutting back on waste generating material and by following clear defined practices of waste management.
  - Create public awareness against big waste generators and provide information to monitor the performance of these sources of waste.
  - Inform the people about waste management program of the government and municipal bodies.
  - Promote public participation in waste management efforts through private partnership where feasible.
  - Propagate the message that the "Clean City Program" is both analytical and purposive and that solutions proposed are within the framework of government initiatives and legally appropriate.
- Citizens co-operation is vital to reduce, reuse and recycling of waste and in keeping garbage streets, by keeping biodegradable "wet" kitchen and food wastes unmixed and separate from recyclable "dry" wastes and other hazardous wastes. Their participation in primary collection of waste, using community bins for storage of waste generated in multi-storeyed buildings, societies, commercial complexes and slums is also essential. If the reasons for doing so are explained, public participation is bound to improve.

4.11 MEASURES TO BE TAKEN TO BRING ABOUT A CHANGE IN PUBLIC (W-15)

- A series of measures can be taken to bring about a change in public behaviour through public awareness programs, which could be as under:

4.11.1 Promote "Reduce, Re-use and Re-cycle (R-R-R)" of Waste (S-18)

4.11.1.1 Reduce

- Everyone is concerned with the growing problems of waste disposal in urban areas with the scarce availability of land for processing and disposal of waste, and environmental remediation measures becoming ever more expensive. It is therefore necessary to not only think about effective ways and means to process and dispose off the waste that we generate each day.
- It is also essential to seriously consider how to avoid or reduce the generation of waste in the first place and to consider ways to re-use and recycle the waste, so that the least quantity of waste needs to be processed and disposed off.
- While the quantity of food waste generated per capita has remained almost static, the quantity of packaging waste material and non-bio-degradable waste is going up alarmingly every year. This increases the burden on local bodies to deal with the problem of non-biodegradable and non-recyclable components of waste landing up at processing and disposal sites.
- The following measures are therefore proposed to be taken to Reduce, Reuse and Recycling of waste by all concerned:
  - All manufacturers producing a variety of domestic and non-domestic products, food as well as non-food should be persuaded to seriously endeavour to use re-usable packaging materials so that after the delivery of goods, the packaging materials could be collected back and used over and over again. They could also consider minimizing or avoiding use of unnecessary packaging materials by innovative methods.
  - Incentives and product discount should be given to consumers for the return of packaging or bottling materials in good condition, to the waste producers or retailers to promote re-use.
  - The cost of packed articles and article without the packaging material could be kept different with a choice to the consumers to take the article without the packaging material at low cost.

4.11.1.2 Re-use

- One person's waste material can be useful material for others. Efforts should therefore be made to encourage collection of such re-usable material through waste collectors, waste producers, NGOs and private sector, instead of allowing reusable waste to land upon the disposal sites. Bottles, cans, tins, drums and cartons can be reused.

4.11.1.3 Re-Cycling

- In the age of excessive packaging materials being used, a lot of recyclable waste material is generated. All-out efforts are necessary to retrieve recyclable material from the households, shops and establishments and fed to the recycling industries through intermediaries such as waste purchasers, waste collectors/ NGOs etc.

4.11.2 Promote Public Participation in SWM Systems Adopted

- The first and foremost thing that the citizens need to be told and made to understand is that no waste shall be thrown on the streets, drains, water bodies, open spaces etc. and that they should form habit of:
  - Storage of wet food/bio-degradable waste and dry recyclable waste separately at source.
  - Participation in primary collection of wastes.
  - Handing over of recyclable waste materials to rag pickers/waste collectors.
  - Use of community bins wherever directed/provided.
  - Use of litter-bins on roads and public places.

4.11.3 Provide Information Hot-line

- The key to success of any public-education, awareness and motivation program is to provide as many ways as possible for the public to interact, as promptly and conveniently as possible, with policy-makers, to seek clarification of doubts, share ideas or give suggestions which are constructively followed up.
- A telephone hot line or Post Box number for written communications could be one of the ways to have inputs from members of the public. The phone must be attended during working hours by polite, responsive and dynamic persons who are well informed, interested in the subject and available at all stated times.

4.11.4 Public Education

- The communication material developed should be utilised in public awareness programmes through variety of approaches as under.

4.11.4.1 Group Education

- This may be done through:

  (a) Group Meetings in the community
  (b) Workshops
  (c) Exhibitions
  (d) Lecture series
  (e) Panel Discussions etc.

4.11.4.2 Mass Education

- This is very essential to cover the entire population as it is not possible to reach all the people through group education programs.

Role of Mass Education in SWM: **(S-15)**

- Advertisement may be given in a planned manner to educate the masses, and local newspapers can also be requested to insert the given messages on SWM at regular intervals.
- Slides in cinema theaters can be displayed to inform and motivate the public.

- Use of TV/ Cable TV/ Radio/Websites is very powerful medium and can be used through local programs to inform the citizen about new waste collection arrangement by local bodies.
- Attractive posters with good photographs and messages with a very few words, should be prepared and displayed in various parts of the city.
- Newspapers may be especially encouraged to give coverage to successful initiatives that have overcome SWM problems.

Importance of Mass Education in Solid Waste Management: **(W-16)**

The mass education in SWM is very important because of following points:

1. To increase the awareness of planning, importance and significance of solid waste management among the people.
2. To increase the efficiency and effectiveness of planning process and implementation of solid waste management.
3. To improve the waste management strategies, negotiations with municipal authorities for better involvement in decision making.
4. To achieve the 3R principles.
5. To play an important role in the permitting process in case of hazardous waste as well as municipal waste facilities.
6. To reduce littering of waste on streets and into drains, open spaces etc.
7. To encourage and assist the local composting and recycling initiatives.

- Mass Education programs can be planned using following methods of communication.

(i) Use of Print Media:

- Advertisements may be given in a planned manner to educate the masses, and local newspapers can also be requested to insert the given messages on SWM at regular intervals. They should also be encouraged to start a regular suggestion.
- Box from where good ideas can be picked up by the local body.
- Newspapers may be specially encouraged to give coverage to successful initiatives that have overcome SWM problems.
- ULBs can also use newspaper delivery services by inserting handbills for readers in a particular locality to announce the start of campaign from time to time and to adhere to the systems introduced.

(ii) Use of TV / Cable TV / Radio/Web Site:

- This is the very powerful medium and can be used through local programs to inform the citizens about new waste collection arrangements made by the local body as and when they become operational and advise them to participate effectively in the prescribed manner. Contact numbers of the concerned officials for problem solving or reporting of SWM grievances may also be publicized.
- This media may be used to publicize successful efforts in some localities to motivate other citizens to perform likewise and get similar recognition of their effort.

(iii) Use of Cinema Halls:

- Slides in cinema theaters can be displayed to inform and motivate the public.

(iv) Street Plays, Puppet Shows etc.:

- Street plays and puppet shows play a significant role in bringing awareness among the people. This method of communication will work well in low-income population; more particularly in slums. Well designed street plays /puppet shows can convey the messages effectively as such programs are well attended in slums.

(v) Posters:

- Attractive posters with good photographs and messages with a very few words, readable from a distance, should be prepared and displayed in various parts of the city where awareness campaign is being taken up.

(vi) Pamphlets:

- Pamphlets, hand bills can be printed giving instructions in very simple and understandable language showing photographs in action and circulated in the community requesting public participation.

(vii) Use of Hoarding:

- Special hoarding may be put at strategic locations in the city carrying messages seeking public participation. Alternatively, all Municipal-licensed hoarding should have a space reserved at the bottom for civic messages. Such messages should be developed and painted by professional agencies. These hoardings should also carry the contact numbers etc.

(viii) Use of Public Transport System:

- Brief messages can be painted on the rear of public buses or inside the bus panels. Public and private firms having their own bus fleets may be invited to support such efforts.

(ix) Use of School Children:

- Children are powerful communicators. Parents who do not listen to the advice of others often take their children seriously. Children are idealistic and would like to change their world for better. The ULB should hold regular meetings with principals, teachers and students to explain the need for change, and the usefulness to society of new ways to manage waste.
- The message can be reinforced by holding essay, debate or drawing and painting competitions on the subject and publicizing the winning contestants. Social clubs can be encouraged to sponsor such events to keep the topic alive. The leading schools could be persuaded to work as a role model for other schools in taking up awareness campaigns in the city through their students, which should be highly publicized and other schools could be persuaded to follow suite.

(x) Primary School Curriculum to Cover the Subject:

- It is an established fact that people form habit at a very young age and habits are hard to change. It is, therefore, necessary to educate young children when they are in primary school to form good habits for managing waste. School curriculum should cover this aspect in the subject of moral science or social studies. This will go a long way in developing enlightened community and least efforts would be required to discipline the people in managing the waste.

(xi) Involvement of National Cadet Corps (NCC), National Social Service (NSS) and Scouts:

- In the schools and colleges the students are participating in NCC, NSS and Scout activities. These students could also be sensitized on the public participatory aspect in solid waste management and as part of their activities they can be involved in the awareness campaign to bring about a change in public behaviour.

(xii) Involvement of Religious Leaders:

- Religious leaders play a significant role in bringing about a change in the mind set of the people. If they advise their devotees/disciples to keep their surroundings clean by not littering anywhere and by managing their waste as advised by the urban local body it will go a long way in improving the situation in the urban areas.

(xiii) Involvement of Medical Practitioners:

- Medical practitioners are held in high esteem by the citizens. A word from them to the patients or the community to practice appropriate systems of waste management at home, offices, shops and establishments would help substantially in bringing compliance of the directions of the urban local body to keep the city clean.

(xiv) Involvement of Mahila Mandals/Women Associations:

- Women are generally found more concerned in maintenance of health and hygiene and they are involved in domestic waste management on day to day basis.
- The awareness among the women could be raised through Mahila Mandals/Women Associations who could be given talking points and necessary literatures in a very simple understandable language/graphics for creating awareness among women.

(xv) Resident Associations:

- Most citizens want a nearby facility to dispose of their waste, but nobody wants a dustbin at their doorstep. Both needs can be met by the house-to-house collection system through handcarts or tricycles. Neighbourhoods can be rewarded for good response to doorstep collection of segregated waste. Groups that undertake to manage the cleaning of their own area can be rewarded by ULBs through grants/subsidies.

(xvi) Voluntary Organizations/NGO Involvement:

- Many NGOs are committed to improve SWM practices in urban areas to protect the environment and have been very active in this field. They have also developed good mass-communication skills and education programs for the public. Such NGOs may be persuaded to actively support the new strategies adopted by the local body and associate in public awareness campaigns. Those who wish to conduct programs for sections of the public on the new SWM strategies may be encouraged to do so and given necessary support.

Marketing Strategies for Recyclable:

There are a number of techniques, recycling business could use to reach potential new customers. These include:

1. Mailshots by post or email. If you use bought business mailing lists, it is important to ensure that these are as up-to-date as possible with valid contact details for key decision makers.
2. Advertising in the press, radio, billboards and even television is useful in terms of raising awareness services.
3. Public relations and press releases may help to increase awareness, but may not lead to many new enquiries. Collection vehicle branding to promote services. This is relatively cheap and is important in presenting a professional image to current and potential customers.
4. Online marketing via a dedicated website or search engines advertisements can be done. A well-designed website can be a relatively low-cost way of attracting new customers.

4.11.5 Role of Civil Engineer to Promote Four R (4R) Principle of SWM (S-15)

A civil engineer can promote 4R principle of SWM by considering following points:

1. **Reduction:** Waste reduction can be achieved in three ways:
   (i) Reducing the amount of material used per product without sacrificing the utility of that product.
   (ii) Increase lifetime of a product.
   (iii) Eliminating the need of the product.

   To reduce waste we usually have to make significant lifestyle changes. Reduce office paper waste by implementing a formal policy to duplex all draft reports and by making training manuals and personnel information available electronically. Improve product design to use less material. Switch to reusable transport containers.
2. **Reuse:** Reuse corrugated moving boxes internally. Encourage employees to reuse office materials rather than purchase new ones.
3. **Recycle:** Recycling turns the material that would otherwise become waste into valuable resources and generates a combination of environmental, financial and social benefits. After collection of these materials (e.g. Glass, metal, plastics, paper etc.) are separated and sent to facilitates that can process them into new products and materials.
4. **Recovery:** Recovery of solid waste means recovering material from solid waste to useful purposes. Composting is a way to return nutrient back into the environment by allowing micro organisms to turn the waste into manure.

Do's and Don'ts to Avoid Health Problem During Solid Waste Management: (W-18)

Do's and Don'ts to avoid health problems during solid waste management are as follows.

Do's:

1. Frequent and regular collection and disposal of municipal solid waste.
2. Separate collection of Non hazardous Domestic and hazardous Industrial Wastes.
3. Safety precautions including use of hand gloves, masks, gum boots etc. during solid waste handling.
4. Regular health check up capms for workers and public residing near dumping yards.

Don't's:

1. Burning of solid waste in open yards, plots.
2. Mixing of Dry and Wet solid Domestic Wastes.
3. Discharging industrial liquid waste in natural drains.
4. Mixing of recyclables in municipal solid waste.

How Enforcement can Help in Stopping Littering in the street ?

1. Penalty provisions for littering and non-compliance in the door to door collection system through Bye laws.
2. The first solution to littering is to always make sure you properly dispose off your garbage.
3. Recycle all that can be recycled and put all other garbage in the proper containers.
4. If someone throwing something on the ground, ask them nicely to put in the trash receptacle instead of on the ground.
5. Responsibilities of resident and businesses for setting out trash in a responsible manner such that trash/ recyclables do not become littered.
6. Responsibility of haulers to collect with care and ensure loads are properly covered to avoid scattering of materials.
7. Property owners must ensure that waste is not stored /stock piled on property for more than a specified period.

4.12 ENFORCEMENT

- All said and done, all human beings are not the same. There are people who understand easily as soon as they are told to behave, there are also people who are hard to understand and there is a special category of people who do not want to understand.
- While all efforts should be made to educate the people to effectively participate in the management of waste, they also need to be told that they can be punished if they fail to discharge their civic duties. The provision of penalties may be made known to the people and details of those punished should be publicized widely to deter others.
- To begin with, the enforcement should begin at the public places, market places, etc. and gradually extended to cover residential areas. Discipline should be brought about in the public offices first so that correct examples be set before the people.

Important Points

- Medical care is vital for our life, health and wellbeing. But the waste generated from medical activities can be hazardous, toxic and even lethal because of their high potential for diseases transmission.
- Safe disposal of biomedical waste is now a legal requirement in India. The ministry of Environment and Forests notified the Biomedical Waste (Management and Handling) Rules, 1998 in July 1998.

- Waste types not to be incinerated are:
  - Pressurized gas containers.
  - Large amount of reactive chemical wastes.
  - Silver salts and photographic or radiographic wastes.
  - Halogenated plastics such as PVC.
  - Waste with high mercury or cadmium content such as broken thermometers, used batteries.
  - Sealed ampoules or ampoules containing heavy metals.
- There are potential risks to environment and health from improper handling of solid wastes. Direct health risks concern mainly the workers in this field, who need to be protected, as far as possible, from contact with wastes.
- **Types of Environmental and Health Hazards:**
  - Chemical
  - Physical
  - Biological
  - Psychological
- Public is in the centre of all the activities, yet it is ignored by the decision makers and made to merely wait and watch and ultimately what people get in hand is what they do not want or what is not in their priority. This creates a gap between the administrators and those administered and an atmosphere of apathy is created which distances people from government initiatives.
- Public involvement can be increased by public awareness programs, effective community participation, transparent and clean administration, introduction of citizen chapters and accountability at all levels.
- A series of measures can be taken to bring about a change in public behaviour through public awareness programs.
- In the age of excessive packaging materials being used, a lot of recyclable waste material is generated. All-out efforts are necessary to retrieve recyclable material from the households, shops and establishments and fed to the recycling industries through intermediaries such as waste purchasers, waste collectors/ NGOs etc.

Practice Questions

1. Define biomedical waste.
2. Give sources of generation of biomedical waste.
3. Give classification of biomedical waste.
4. State different components of biomedical waste.
5. State biomedical waste technology.
6. Describe disposal of biomedical waste.
7. Give effects of biomedical waste.
8. Enlist types of environmental health hazards.
9. Explain health problem during time of segregation of solid waste.
10. State different methods of communication for mass education.
11. Describe the importance of public participation in SWM.
12. State the importance of mass education in solid waste management.

MSBTE Questions and Answers (As Per 'G' Scheme)

Summer 2016

1. What are the sources and generation of biomedical waste? Give its health aspects. **(4M)**

Ans. Refer Sections 4.2, 4.3 and 4.7.

2. Describe Biomedical waste management and handling as per rule 1998. **(4M)**

Ans. Refer Sections 4.2 and 4.4.

3. State the classification of Biomedical waste. **(4M)**

Ans. Refer Section 4.3.

4. State the importance of public involvement in solid waste management. **(4M)**

Ans. Refer Section 4.9.

5. Explain the health problem during the time of segregation, reuse, recovery recycling of solid waste. **(4M)**

Ans. Refer Section 4.7.

Winter 2016

1. Define Biomedical Waste and enlist any four sources of generation of biomedical waste. **(4M)**

Ans. Refer Sections 4.1.1 and 4.2.

2. Explain the health problem during time of segregation of solid waste. **(4M)**

Ans. Refer Section 4.7.

3. Explain Biomedical waste management and handling Rule 1998. **(4M)**

Ans. Refer Sections 4.4 and 4.5.

4. Write short note on health aspect during handling and processing of solid waste. **(4M)**

Ans. Refer Sections 4.7 and 2.2.5.

5. Describe health problems during time of segregation of solid waste. **(4M)**

Ans. Refer Section 4.7.

6. State the importance of mass education in solid waste management. **(4M)**

Ans. Refer Section 4.11.4.2.

Summer 2017

1. Explain the health safety measures to be adopted during handling and processing solid waste. **(4M)**

Ans. Refer Sections 4.7 and 2.2.5.

2. Explain four salient features mentioned in biomedical waste management and handling as per rule 1998. **(4M)**

Ans. Refer Section 4.4, Table 4.3.

3. Describe method of collection and disposal of biomedical waste. **(8M)**

Ans. Refer Section 4.6.

4. Explain the importance of Public involvement and participation in Solid Waste Management.

Ans. Refer Section 4.9. **(8M)**

Winter 2017

1. Enlist sources of biomedical waste. **(2M)**

Ans. Refer Section 4.2.

2. State any two features of Biomedical Waste Management of Handling Rule, 1998. **(2M)**

Ans. Refer Table 4.4.

3. Enlist any two methods of mass education in SWM. **(2M)**

Ans. Refer Section 4.11.4.2.

4. Classify biomedical waste treatment w.r.t. option, category, treatment and disposal. **(8M)**

Ans. Refer Section 4.3.

5. Define communicable and non-communicable diseases. Discuss the importance of public involvement and participation in SWM. **(8M)**

Ans. Refer Section 4.9.

Summer 2018

1. Explain about any four effects of biomedical waste. **(4M)**

Ans. Refer Section 4.7.1.

2. State about the strategy for public participation in solid waste management. **(4M)**

Ans. Refer Section 4.10.

3. Define - Biomedical waste and write down its components. **(4M)**

Ans. Refer Section 4.1.1.

4. Enlist the varieties of industrial solid waste. Explain in brief. **(4M)**

Ans. Refer Table 4.2.

5. State the health problems that may be faced during segregation, reuse, recovery and recycling of solid waste. **(4M)**

Ans. Refer Section 4.8.

6. State the measures to be taken to bring about a change in public for solid waste management w.r.t. '3R' strategy. **(4M)**

Ans. Refer Section 4.11.1.

7. Explain in detail about Biomedical waste management and handling as per Rule 1998. **(4M)**

Ans. Refer Section 4.4, Table 4.3.

Winter 2018

1. State the schedule - II of biomedical waste (Management and Handling) Rules 1998. **(4M)**

Ans. Refer Table 4.3.

2. State sources of bio-medical waste and give average BMW generation rate. **(4M)**

Ans. Refer Table 4.1.

3. State biomedical waste management technologies and explain any one. **(4M)**

Ans. Refer Section 4.2.

4. State hazardous substances in E-waste and its health effects on health and environment. **(4M)**

Ans. Refer Table 4.5.

5. State the importance of public involvement and participation in solid waste management. **(4M)**

Ans. Refer Section 4.9.

6. State Do's and Don'ts to avoid health problem during solid waste management. **(4M)**

Ans. Refer Section 4.11.5.

7. Explain process of public involvement and participation in solid waste management. **(4M)**

Ans. Refer Section 4.9.

Summer 2019

1. Describe the provisions in law for safe disposal of bio-medical waste in India. **(4M)**

Ans. Refer Section 4.4.

2. State different methods of communication for mass education. **(4M)**

Ans. Refer Section 4.11.4.2.

3. Describe the importance of public participation in solid waste management. **(4M)**

Ans. Refer Section 4.9.

4. State different components of biomedical waste. **(4M)**

Ans. Refer Section 4.1.2.

5. Describe the biomedical waste management techniques. **(4M)**

Ans. Refer Section 4.6.

***

Chapter **5...**

Industrial Waste Management and E-waste Management

Weightage of Marks = 14, Teaching Hours = 08

Syllabus

Industrial Waste Management :

5.1 Variety of Industrial Waste

5.2 Collection and Disposal of Industrial Waste

5.3 Control Measures for Industrial Waste

5.4 Recycling of Industrial Waste

E-waste Management :

5.5 Definition of E-waste, Varieties of E-wastes, Dangers of E-waste

5.6 Recycling of E-waste

5.7 Disposal of E-waste

About this Chapter

At the end of this chapter, students will be able to:

- Explain the relevant method of disposal of given type of industrial waste.
- Classify the industrial waste based on the given criteria.
- Describe the process of controlling the generation of given type of industrial waste at the given site.
- Suggest the relevant appliance for the disposal of given type of E-waste with justification.
- Explain the ill effects of given type of E-waste on the environment of that area.
- Suggest the relevant method of recycling and disposal of the given type of E-waste in the given situation.

5.0 OVERVIEW

- Environmental pollution is the major problem associated with rapid industrialisation, urbanisation and rise in living standards of people. For developing countries, industrialisation was must and still this activity very much demands to build self-reliant and plays a major role also in uplifting nation's economy.
- However, industrialisation on the other hand, has also caused serious problems relating to environmental pollution. Therefore, wastes seem to be a by-product of growth. The country like India can not afford to lose them as sheer waste.
- On the other hand, with increasing demand for raw materials for industrial production, the non-renewable resources are dwindling day-by-day. Therefore, efforts are to be made for controlling pollution arising out of the disposal of wastes by conversion of these unwanted wastes into utilisable raw materials for various beneficial uses.

- The problems relating to disposal of industrial solid waste are associated with lack of infrastructural facilities and negligence of industries to take proper safeguards. In number of cities and towns, small scale industries find it easy to dispose waste here and there and it makes difficult for local bodies to collect such waste though it is not their responsibility.
- In some cities, industrial, residential and commercial areas are mixed and thus all waste gets intermingled. Therefore, it becomes necessary that the local bodies along with State Pollution Control Board (SPCB) work out requisite strategy for organising proper collection and disposal of industrial solid waste.
- Management of Industrial Solid Waste (ISW) is not the responsibility of local bodies. Industries generating solid waste have to manage such waste by themselves and are required to seek authorisations from respective State Pollution Control Boards (SPCBs) under relevant rules. However, through joint efforts of SPCBs, local bodies and the industries, a mechanism could be evolved for better management.

5.1 VARIETY OF INDUSTRIAL SOLID WASTE

- The major generators of industrial solid wastes are the thermal power plants producing coal ash, the integrated Iron and Steel mills producing blast furnace slag and steel melting slag, non-ferrous industries like aluminium, zinc and copper producing red mud and tailings, sugar industries generating press mud, pulp and paper industries producing lime and fertilizer and allied industries producing gypsum.

  Examples of industrial wastes are: **(W-15)**

  - Chemical solvents
  - Paints
  - Sandpaper
  - Paper products
  - Industrial by-products
  - Metals
  - Radioactive wastes
  - Red mud
  - Coal ash

5.1.1 Coal Ash

- In general, a 1,000 MW station using coal of 3,500 kCal/kg and ash content in the range of 40 - 50% would need about 500 hectares for disposal of fly ash for about 30 years' operation. It is, therefore, necessary that fly ash should be utilised wherever possible to minimize environmental degradation.
- The thermal power plant should take into account the capital and operation/maintenance cost of fly ash disposal system as well as the associated environmental protection cost, as compared with dry system of collection and its utilisation by the thermal power plant or other industry, in evaluating the feasibility of such system.
- The research and development carried out in India for utilisation of fly ash for making building materials has proved that fly ash can be successfully utilised for production of bricks, cement and other building materials. Local technology for construction of building materials utilising fly ash is available and are being practised in a few industries. However, large scale utilisation is yet to take off.
- Even if the full potential of fly ash utilisation through manufacture of fly ash bricks and blocks is explored, the quantity of fly ash produced by the thermal power plants are so huge that major portion of it will still remain unutilised. Hence, there is a need to evolve strategies and plans for safe and environmentally sound method of disposal.

5.1.2 Integrated Iron and Steel Plant Slag

- The Blast Furnace (BF) and Steel Melting Shop (SMS) slags in integrated iron and steel plants are at present dumped in the surrounding areas of the steel plants making hillocks encroaching on the agricultural land.
- Although, the BF slag has potential for conversion into granulated slag, which is a useful raw material in cement manufacturing, it is yet to be practised in a big way. Even the use of slag as road subgrade or land-filling is also very limited.

5.1.3 Phosphogypsum

- Phosphogypsum is the waste generated from the phosphoric acid, ammonium phosphate and hydrofluoric acid plants. This is very useful as a building material.
- At present, very little attention has been paid to its utilisation in making cement, gypsum board, partition panel, ceiling tiles, artificial marble, fiber boards etc.

5.1.4 Red Mud

- Red mud as solid waste is generated in non-ferrous metal extraction industries like aluminium and copper. The red mud at present is disposed in tailing ponds for settling, which more often than not finds its course into the rivers, especially during monsoon. However, red mud has recently been successfully tried and a plant has been set up in the country for making corrugated sheets.
- Demand for such sheet should be popularised and encouraged for use. This may replace asbestos which is imported and also banned in developed countries for its hazardous effect. Attempts are also made to manufacture polymer and natural fibres composite panel doors from red mud.

5.1.5 Lime Mud

- Lime sludge, also known as lime mud, is generated in pulp and paper mills which is not recovered for reclamation of calcium oxide for use except in the large mills. The lime mud disposal by dumping into low-lying areas or into water courses directly or as run-off during monsoon is not only creating serious pollution problem but also wasting the valuable non-renewable resources.
- The reasons for not reclaiming the calcium oxide in the sludge after recalcination is that, it contains high amount of silica. Although a few technologies have been developed to desilicate black liquor before burning, none of the mills in the country are adopting desilication technology.

5.1.6 Waste Sludge and Residues

- Treatment of industrial wastes/effluents results in generation of waste sludge/ residues which, if not properly disposed, may cause ground and surface water pollution.

5.1.7 Potential Reuse of Solid Wastes

- Research and Development (R&D) studies conducted by the R&D Institutions like Central Building Research Institute, Roorkee (CBRI) and the National Council for Building Research, Ballabgarh (NCBR) reveal that the aforesaid solid wastes has a very good potential to be utilised in the manufacture of various building materials.

5.2 COLLECTION, STORAGE, TREATMENT AND DISPOSAL OF INDUSTRIAL SOLID WASTE (W-18)

5.2.1 Waste Segregation

- Many wastes are mixtures of hazardous and non-hazardous wastes. Much of their contents may even be water. By segregating key toxic constituents, isolating liquid fraction, keeping hazardous streams away from non-hazardous wastes, generator can save substantial amounts of money on disposal or find new opportunities for recycling and reuse of wastes.
- The Ministry of Environment, Government of India, had identified toxicity of different chemicals, through the 'Manufacture, Storage and Import of Hazardous Chemicals Rules, 1989' in exercise of power conferred by Section 6, 8 and 25 of Environment Protection (E.P). Act, 1986, and had notified mandatory requirements for its management.
- In India, quantum of generation of wastes (solid/liquid and hazardous/non-hazardous) for different industries has not been detailed, which is necessary for wastes exchange system or for adopting treatment/disposal alternatives for different wastes segregated.

5.2.2 Collection, Storage and Transport

- The unsatisfactory state of storage of hazardous wastes can be remedied to a large degree by such low-cost measures as restricting access, fencing off the storage area to minimise any wind-blown nuisance, providing separate covered storage for putrifiable of hazardous wastes, and ensuring regular and frequent collection.
- There are certain measures a municipal authority can take to control the transportation of industrial wastes, even if it does not want to become actually involved itself. For example, contractors should be licensed after ensuring that they are technically competent and environmentally aware and should be allowed to handle industrial wastes.
- Labelling and coding of hazardous waste load can be made mandatory so that in the event of an accident, the emergency services know how to handle a spillage.
- Municipal authorities can be given the responsibility to monitor the contractors to minimise cases of fly-tipping and ensure that industrial wastes are disposed at the appropriate sites.
- If a municipal authority can also collect industrial waste; industries must pay the charge which will be based on the quantity and nature of the waste.
- This might minimise the quantity of waste produced by industry and at the same time the programme will become financially viable and self-sustaining. The principle 'the polluter pays' should be adhered to in all such cases.

5.2.3 Disposal Methods (W-16)

- Depending upon the characteristics of the wastes, different types of disposal methods can be used for hazardous and non-hazardous industrial wastes. The most predominant and widely practised methods for wastes disposal are: (a) Landfill, (b) Incineration and (c) Composting. These methods are already described in Chapter no. 3 in detail.

Method of Collection and Disposal of Industrial Waste: (W-16, 18)

- Industrial waste may be hazardous and non-hazardous and collection methods will differ accordingly. This is done by industry itself or sometimes, it is possible that they pay charges according to quantity and nature, to municipal authority and the persons in municipal corporation collect the industrial waste.
- The collection of industrial waste should be regular and frequent from various industries. The storage of hazardous industrial waste should be in a separate covered shed having proper fencing and restricted access. Technically competent contractors should be allowed to handle industrial waste with environmental awareness. Labeling, coding of hazardous waste with assurance of emergency services in case of spillage during collection and transportation should be made mandatory. The most predominant and practiced methods for industrial waste disposal are Secure Landfill, Incineration and Recycling.
- The disposal of industrial wastes can be used depending upon the characteristics of industrial wastes. The industrial wastes can be categorized as hazardous and non-hazardous waste.
- Recycling method used for disposal of ash in the manufacture of cement, construction materials. Incineration with off gas monitoring method is used for industrial wastes such as solvents, residues etc. Secured land filling method is used for STP sludge, plant sludge etc.

Some of disposal methods in respective industries are given as follows.

Table 5.1

Sr. No.	Industry	Disposal Method
1.	Food industry	Waste is treated either by vermin-composting (for small quantity) and bio-methenation process (for large quantity)
2.	Paper and pulp industry	Paper cutting waste is recycled for either same industry or it is send to cardboard industry as raw material.

Sr. No.	Industry	Disposal Method
3.	Thermal power station industry	Fly ash produced can be recycled • For raw material in ordinary Portland cement. • For cellular concrete bricks and blocks lime and cement fly ash concrete. • To use as a plasticizers.
4.	Sugar industry	Waste is converted into manure by composting treatment. And the baggage is used as fuel for boiler.
5.	Blast furnace slags from metal industry	• As a aggregate in concrete. • Manufacture of different types of cement like slag cement, super sulphated cement, metallurgical cement and non-portland cement. • As a structural fill (air-cooled slag).

5.3 CONTROL MEASURES OF INDUSTRIAL WASTE (S-16, 18)

- A two-stage approach should be thought of for waste management, e.g., (a) prevention and (b) control of environmental pollution. Prevention aims at minimisation of industrial wastes at source, while the latter stresses on treatment and disposal of wastes.

5.3.1 Prevention - A Waste Minimisation Approach

- Reduction and recycling of wastes are generally site/plant specific. Usually, waste minimisation techniques can be grouped into four major categories which are applicable for hazardous as well as non-hazardous wastes. These groups are as follows:

1. **Inventory Management and Improved Operations:**
   - Inventorisation and tracing of all raw materials,
   - Purchasing of fewer toxic and more non-toxic production materials,
   - Implementation of employees' training and management feedback, and
   - Improving material receiving, storage and handling practices.
2. **Modification of Equipment:**
   - Installation of equipment that produce minimal or no wastes,
   - Modification of equipment to enhance recovery or recycling options,
   - Redesigning of equipment or production lines to produce less waste,
   - Improving operating efficiency of equipment, and
   - Maintaining strict preventive maintenance programme.
3. **Production Process Changes:**
   - Substitution of non-hazardous for hazardous raw materials,
   - Segregation of wastes by type for recovery,
   - Elimination of sources of leaks and spills,
   - Separation of hazardous from non-hazardous wastes,
   - Redesigning or reformulation for products to be less hazardous, and
   - Optimisation of reactions and raw material use.
4. **Recycling and Reuse:**
   - Installation of closed-loop systems,
   - Recycling of wastes, and
   - Exchange of wastes.

5.3.2 Waste Management at Source

- *Waste management at source* means the reduction of volume of raw materials and the waste, both, by some means.
- A specific example worth-mentioning in this context is removal of fly ash through coal beneficiation process at the mine head in view of high ash content.
- *Coal beneficiation* is 'a process by which the quality of raw coal is improved by either reducing the extraneous matter that gets extracted along with the mined coal or reducing the associated ash or both'.
- It is evident that the larger the volume of waste and the longer the distance of transportation of raw material (coal), the bigger will be the economic benefit in favour of coal beneficiation instead of carrying the filthy fly ash.
- However, benefit-cost analysis have to be made before taking appropriate decision.

5.3.3 Reduction in Waste Generation

- It is possible to cut down waste generation at source by simple, inexpensive measures modifying production processes, through changes in raw materials/ product design and by employing recovery/recycling and reuse techniques.

5.3.4 Reuse and Recycle

- To avoid treatment, it is important from the environmental pollution view point as well as for the benefit of entrepreneurs to recycle and reuse the wastes generated by adoption of certain process change or by use of low/no-waste generation technology.

5.3.5 Waste Minimisation (W-15)

- Waste minimisation can be practised at various places in the industrial processes. Waste minimisation requires careful planning, creative problem evolving, changing in attitude, sometimes capital investment, and most important a real commitment.
- Mostly, investment on waste minimisation and recovery pays off substantially within a short time.

Waste Minimization Approach for Treating Industrial Solid Waste: (S-15)

- **Optimization of resources:** Waste reduction at individual and institutional level goes side by side with the utilization of raw materials.
- **Using again the scrap material:** This is the process in which individual and industry reuse the waste material as soon as it is produced. This keeps it from becoming a waste material.
- **Quality control improvement and process monitoring:** This technique is to ensure that products produced are kept from rejection and this is increased by the inspection of frequency and monitoring point's inspection.
- **Exchanging waste:** This is the technique in which the waste product, which comes out of a process, becomes a raw material for another process. This is another way for reducing waste.
- **Zero waste:** This is a whole systems approach that aims to eliminate waste at the source and at all points down the supply chain, with the intention of producing no waste. It is a design philosophy which emphasizes waste prevention as opposed to end of pipe waste management.

5.3.6 Initial Investment

- The initial investment for a pollution prevention project may be higher in some cases than the cost of installing conventional pollution control equipment.
- However, the annual operation and maintenance cost of the removal will almost always make the total cost of treatment higher than the total cost of preventive measures at sources.
- However, treatment and disposal of residual waste even after taking preventive measures should be given due consideration.

5.4 RECYCLING OF INDUSTRIAL WASTE (W-15, 16, 18)

Table 5.2

Sr. No.	Waste	Areas of Recycle/Application
1.	Fly ash	(i) Cement. (ii) Raw material in Ordinary Portland Cement (OPC) manufacture. (iii) Manufacture of oilwell cement. (iv) Making sintered fly ash light-weight aggregates. (v) Cement/silicate bonded fly ash/clay binding bricks and insulating bricks. (vi) Cellular concrete bricks and blocks, lime and cement fly ash concrete. (vii) Precast flyash concrete building units. (viii) Structural fill for roads, construction on sites, land reclamation etc. (ix) As filler in mines, in bituminous concrete. (x) As plasticizer. (xi) As water reducer in concrete and sulphate resisting concrete. (xii) Amendment and stabilisation of soil.
2.	Blast Furnace Slags	(i) Manufacture of slag cement, super sulphated cement, metallurgical cement. (ii) Non-portland cement. (iii) Making expansive cement, oilwell, coloured cement and high early-strength cement. (iv) In refractory and in ceramic as sital. (v) As a structural fill (air-cooled slag). (vi) As aggregates in concrete.
3.	Ferro-alloy and other metallurgical slags.	(i) As structural fill. (ii) In making pozzolana metallurgical cement.
4.	Byproduct gypsum	(i) In making of gypsum plaster, plaster boards and slotted tiles. (ii) As set controller in the manufacture of portland cement. (iii) In the manufacture of expensive or non-shrinking cement, super sulphated and anhydrite cement. (iv) As mineraliser. (v) Simultaneous manufacture of cement and sulphuric acid.
5.	Lime sludge (phosphor-chalk, paper and sugar sludges)	(i) As a sweetener for lime in cement manufacture. (ii) Manufacture of lime pozzolana bricks/ binders. (iii) For recycling in parent industry. (iv) Manufacture of building lime. (v) Manufacture of masonry cement.
6.	Chromium sludge	(i) As a raw material component in cement manufacture. (ii) Manufacture of coloured cement as a chromium-bearing material.
7.	Red mud	(i) As a corrective material. (ii) As a binder. (iii) Making construction blocks. (iv) As a cellular concrete additive. (v) Coloured composition for concrete. (vi) Making heavy clay products and red mud bricks. (vii) In the formation of aggregate. (viii) In making floor and all tiles. (ix) Red mud polymer door.
8.	Pulp and Paper	(i) Lignin

(**Source:** Industrial Waste Management, National Waste Management Council (NWMC), 1990).

5.5 E-WASTE (S-16)

5.5.1 Definition (S-16; W-15, 16)

- "Electronic waste" may be defined as 'discarded computers, office electronic equipment, entertainment electronic devices, mobile phones, television sets and refrigerators'. This definition includes used electronics and electrical items which are destined for reuse, resale, salvage, recycling or disposal.
- E-waste or electronic waste, therefore, broadly describes the discarded, surplus, obsolete, broken, electrical or electronic devices.

5.5.2 Composition of E-waste

- E-waste consists of all waste from electronic and electrical appliances which have reached their end-of-life period or are no longer fit for their original intended use and are destined for recovery, recycling or disposal.
- It includes computer and its accessories, monitors, printers, keyboards, central processing units; typewriters, mobile phones and chargers, remotes, compact discs, headphones, batteries, LCD/Plasma TVs, air conditioners, refrigerators and other household appliances.
- The composition of e-waste is diverse and falls under 'hazardous' and 'non-hazardous' categories. Broadly, it consists of ferrous and non-ferrous metals, plastics, glass, wood and plywood, printed circuit boards, concrete, ceramics, rubber and other items.
- Iron and steel constitute about 50% of the waste, followed by plastics (21%), non-ferrous metals (13%) and other constituents. Non-ferrous metals consist of metals like copper, aluminium and precious metals like silver, gold, platinum, palladium and so on.
- The presence of elements like lead, mercury, arsenic, cadmium, selenium, hexavalent chromium, and flame retardants beyond threshold quantities make e-waste hazardous in nature.
- It contains over 1000 different substances, many of which are toxic and creates serious pollution upon disposal. Obsolete computers pose the most significant environmental and health hazard among the e-wastes.

5.5.3 E-waste Generation In India

- All over the world, the quantity of electrical and electronic waste generated each year, especially computers and televisions, has assumed alarming proportions. In 2006, the International Association of Electronics Recyclers (IAER) projected that 3 billion electronic and electrical appliances would become e-waste by 2010. That would be equivalent to an average e-waste generation rate of 400 million units a year till 2010.
- Globally, about 20-50 MT (million tonnes) of e-wastes are disposed off each year, which accounts for 5% of all municipal solid waste. Although no definite official data exist on how much waste is generated in India or how much is disposed off, there are estimations based on independent studies conducted by the NGOs or government agencies.
- According to the Comptroller and Auditor-General's (CAG) report, over 7.2 MT of industrial hazardous waste, 4 lakh tonnes of electronic waste, 1.5 MT of plastic waste, 1.7 MT of medical waste, 48 MT of municipal waste are generated in the country annually.
- In 2005, the Central Pollution Control Board (CPCB) estimated India's e-waste at 1.47 lakh tonnes or 0.573 MT per day. A study released by the Electronics Industry Association of India (ELCINA) at the electronics industry expo – "Componex Nepcon 2009" had estimated the total e-waste generation in India at a whopping 4.34 lakh tonnes by end 2009. The CPCB has estimated that it will exceed the 8 lakh tonnes or 0.8 MT mark by 2012-13.
- There are 10 States that contribute to 70% of the total e-waste generated in the country, while 65 cities generate more than 60% of the total e-waste in India. Among the 10 largest e-waste generating States, Maharashtra ranks first followed by Tamil Nadu, Andhra Pradesh, Uttar Pradesh, West Bengal, Delhi, Karnataka, Gujarat, Madhya Pradesh and Punjab.

- Among the top ten cities generating e-waste, Mumbai ranks first followed by Delhi, Bengaluru, Chennai, Kolkata, Ahmedabad, Hyderabad, Pune, Surat and Nagpur.
- The main sources of electronic waste in India are the government, public and private (industrial) sectors, which account for almost 70% of total waste generation.
- The contribution of individual households is relatively small at about 15%; the rest being contributed by manufacturers. Though individual households are not large contributors to waste generated by computers, they consume large quantities of consumer durables and are, therefore, potential creators of waste.
- An Indian Market Research Bureau (IMRB) survey of 'E-waste generation at Source' in 2009 found that out of the total e-waste volume in India, televisions and desktops including servers comprised 68% and 27% respectively. Imports and mobile phones comprised of 2% and 1% respectively.
- As a large-scale organised e-waste recycling facility, the Attero Recycling Plant in Roorkee opened in January 2010. Despite 23 units currently registered with the Government of India, Ministry of Environment and Forests/Central Pollution Control Board, as e-waste recyclers/reprocessors, having environmentally sound management facilities, the entire recycling process more or less still exists in the unorganised sector.
- The Ministry of Environment and Forests (MoEF) has notified the Hazardous Wastes (Management, Handling and Trans-boundary Movement) Rules, 2008 for effective management of hazardous wastes, including e-waste in the country. But these rules do not apply to the radioactive wastes such as Cobalt-60 which are covered under the Atomic Energy Act, 1962.

5.5.4 Varieties of E-waste/Examples of E-waste (S-16)

Table 5.3: Identification of E-waste Categories

Type of E-waste	Examples
1. Large Household Appliances	Washing machines, Dryers, Refrigerators, Air conditioners etc.
2. Small Household Appliances	Vacuum cleaners, Coffee machines, Irons, Toasters etc.
3. Office, Information and Communication Equipment	PC's, Laptops, Mobiles, Telephones, Fax machines, Copiers, Printers etc.
4. Entertainment and Consumer, Electronics and Toys, Leisure, Sports and Recreational Equipment, and Automatic Issuing Machines	Televisions, VCR/DVD/DC players, Hi-Fi sets, Radios etc. and Electric train sets, coin slot machines, treadmills etc. and Vending machines, parking ticket equipment etc.
5. Lighting Equipment	Fluorescent tubes and lamps, sodium lamps, electric fused bulbs and tubelights etc.
6. Electric and Electronic Tools	Drills, Electric saws, Sewing machines, Lawn mowers etc.
7. Security and Health Care Equipment	Surveillance and Control Equipment (e.g. CCTV cameras, scanning equipment), and Medical Instruments and Equipment (e.g. X-ray and heart lung machines) etc.

5.5.5 Dangers of E-waste (S-16)

- Some computer components can be reused in assembling new computer products, while others are reduced to metals that can be reused in applications as varied as construction, flatware and jewelry.
- Substances found in large quantities include epoxy resins, fiberglass, PCBs, PVC (polyvinyl chlorides), thermosetting plastics, lead, tin, copper, silicon, beryllium, carbon, iron and aluminium.

- Elements found in small amounts include cadmium, mercury and thallium.
- Elements found in trace amounts include americium, antimony, arsenic, barium, bismuth, boron, cobalt, europium, gallium, germanium, gold, indium, lithium, manganese, nickel, niobium, palladium, platinum, rhodium, ruthenium, selenium, silver, tantalum, terbium, thorium, titanium, vanadium and yttrium.
- Almost all electronics contain lead and tin (as solder) and copper (as wire and printed circuit board tracks), though the use of lead-free solder is now spreading rapidly.

The following are the effects of substances that come out of E-waste:

Hazardous Substances:

- **Americium:** The radioactive source in smoke alarms. It is known to be carcinogenic.
- **Mercury:** It is found in fluorescent tubes (numerous applications), tilt switches (mechanical doorbells, thermostats), and flat screen monitors. Health effects include sensory impairment, dermatitis, memory loss and muscle weakness. Environmental effects in animals include death, reduced fertility, slower growth and development.
- **Sulphur:** It is found in lead-acid batteries. Health effects include liver damage, kidney damage, heart damage, eye and throat irritation. When released into the environment, it can create sulphuric acid.
- **BFRs:** It is used as flame retardants in plastics in most electronics. Includes PBBs, PBDE, DecaBDE, OctaBDE, PentaBDE. Health effects include impaired development of the nervous system, thyroid problems, liver problems. Environmental effects: Similar effects as in animals as humans. PBBs were banned from 1973 to 1977 and onwards. PCBs were banned during 1980s.
- **Cadmium:** It is found in light-sensitive resistors, corrosion-resistant alloys for marine and aviation environments, and nickel-cadmium batteries. The most common form of cadmium is found in Nickel-cadmium rechargeable batteries. These batteries tend to contain between 6% to 18% cadmium. The sale of Nickel-Cadmium batteries has been banned in the European Union except for medical use. When not properly recycled it can leach into the soil, harming microorganisms and disrupting the soil ecosystem. Exposure is caused by proximity to hazardous waste sites and factories and workers in the metal refining industry. The inhalation of cadmium can cause severe damage to the lungs and is also known to cause kidney damage.
- **Lead:** Solder, CRT monitor glass, lead-acid batteries, some formulations of PVC. A typical 15-inch cathode ray tube may contain 1.5 pounds of lead, but other CRTs have been estimated as having upto 8 pounds of lead.
- **Beryllium oxide:** Filler in some thermal interface materials such as thermal grease used on heat sinks for CPUs and power transistors, magnetrons, X-ray-transparent ceramic windows, heat transfer fins in vacuum tubes and gas lasers.

Non-hazardous Substances:

- Tin: Solder, coatings on component leads.
- Copper: Copper wire, printed circuit board tracks, component leads.
- Aluminium: Nearly all electronic goods using more than a few watts of power (heatsinks), electrolytic capacitors.
- Iron: Steel chassis, cases and fixings.
- Germanium: 1950s–1960s transistorized electronics (bipolar junction transistors).
- Silicon: Glass, transistors, ICs, printed circuit boards.
- Nickel: Nickel-cadmium batteries.
- Lithium: Lithium-ion batteries.
- Zinc: Plating for steel parts.
- Gold: Connector plating, primarily in computer equipment.

Table 5.4: Effects of E-Waste Constituent on Health (S-15; W-15, 18)

Source of e-wastes	Constituent	Health effects
Solder in printed circuit boards, glass panels and gaskets in computer monitors.	Lead (PB)	• Damage to central and peripheral nervous systems, blood systems and kidney damage. • Affects brain development of children.
Chip resistors and semiconductors	Cadmium (CD)	• Toxic irreversible effects on human health. • Accumulates in kidney and liver. • Causes neural damage. • Teratogenic.
Relays and switches, printed circuit boards	Mercury (Hg)	• Chronic damage to the brain. • Respiratory and skin disorders due to bioaccumulation in fishes.
Corrosion protection of untreated and galvanized steel plates, decorator or hardner for steel housings	Hexavalent chromium (Cr) VI	• Asthmatic bronchitis. • DNA damage.
Cabling and computer housing	Plastics including PVC	• Burning produces dioxin. It causes √ Reproductive and developmental problems; √ Immune system damage; √ Interfere with regulatory hormones
Plastic housing of electronic equipments and circuit boards.	Brominated flame retardants (BFR)	• Disrupts endocrine system functions.
Front panel of CRTs	Barium (Ba)	• Short term exposure causes: √ Muscle weakness; √ Damage to heart, liver and spleen.
Motherboard	Beryllium (Be)	• Carcinogenic (lung cancer) • Inhalation of fumes and dust. Causes chronic beryllium disease or beryllicosis. • Skin diseases such as warts.

Effects of E-Waste on Human Health and Environment:

(i) Effects of e-waste on human health:

- Beryllium causing human carcinogens.
- Cadmium causing damage to kidney and bones.
- Lead damage to nervous system.
- Mercury damage to central nervous system.
- Polychlorinated biphenyls damage liver, and may cause cancer and damage to reproductive system etc.

(ii) Effects of e-waste on environment:

- Pollution of ground water.
- Air pollution.
- Acidification of soil.
- E-waste account for 40% lead and 75% metal in land fills etc.

5.6 RECYCLING OF E-WASTE

- To obtain the highest possible result, labour intensive methods are used to completely disassemble and separate items into core materials and components.

  E-Cycle Recovery will NOT:

  Place your equipment in landfill.

  Refurbish, sell or give away your items.

  Ship to where it may be processed under poor health and safety conditions.

Methods of Recycling of Different Materials:

Plastic

- All plastics are sent to a recycler who uses the raw material to manufacture items such as vineyard stakes, fence posts and plastic sleepers.

Metal

- Scrap metals are sent to a recycler. They are placed through a shredder before magnetic systems to separate the ferrous from the non-ferrous materials. The resulting product is used in the manufacture of new steel and other metal products.

Glass

- Cathode Ray Tube (CRT) glass, used in computer monitors and televisions, are a major issue because they incorporate many hazardous materials. Lead is the most prevalent toxic material in CRT glass; it is poisonous to the nervous system and can remain in the human body for years. Tubes in a large CRT monitor can contain upto 4 kg of lead as well as other toxic metals such as phosphor and barium.
- To obtain the highest environmental outcome it is necessary to send the glass to CRT recycling facilities where it will be processed to specification before being used in the manufacture of new CRT monitors and televisions.

Mercury

- Mercury is commonly found within many e-waste items. Highly toxic, even in small amounts, it has been known to cause damage to the lungs, kidneys, brain, nervous and reproductive systems. Given the opportunity to leach into water and soil, it is able to be ingested by aquatic creatures and then through the food chain into our diet.
- To avoid these consequences, one should remove mercury containing devices such as tubes and lamps and forward these to an CPCB approved mercury recycling plant. Here they use technology that captures the mercury for use in dental amalgams, separates the glass for use as glass wool in home insulation and takes out the phosphor powder for use in fertilizer products.

Wood

- Commonly found in older televisions, stereos and speakers. Wood is sent to recycling companies who either shred it for use as much or use innovative technology to mix it with other waste materials for use as an alternative fuel source.

Printed Circuit Boards

- Circuit boards are sent to ISO 14001 (Environmental Management) accredited companies. Here they can be processed in specialised smelters to recover non-renewable resources such as copper, gold, silver, palladium and other precious metals.

Hard Drives

- Hard drives, in whole and shredded form, are sent to an aluminium foundry for processing into aluminium ingots. The majority of aluminium ingots are used within the automotive industry.

Toner and Ink Cartridges

- Toner and Ink cartridges are packaged in a sealed box and returned to industry recyclers. Some will be remanufactured into new cartridges, and the remainder that can not be remanufactured will be separated into plastic and metal and returned to the recycle chain as raw materials.

Batteries

- Nickel Cadmium, Nickel Metal Hydride and Lithium Ion batteries are sent to recycling company who meets the strict environmental standards. The material is hulled to remove excess plastic, and then the metals placed in special smelter pots to recover cobalt, cadmium, nickel and steel for reuse in a range of areas such as battery production and stainless steel fabrication.

CD ROMs, Sound and Memory Cards

- For copyright and security reasons these products are shredded before being sent to plastic and metal recyclers.

5.7 MANAGEMENT OF E-WASTE (W-18)

- It is estimated that 75% of electronic items are stored due to uncertainty of how to manage it. These electronic junks lie unattended in houses, offices, warehouses etc. and normally mixed with household wastes, which are finally disposed off at landfills. This necessitates implementable management measures.
- In industries, management of e-waste should begin at the point of generation. This can be done by waste minimization techniques and by sustainable product design. Waste minimization in industries involves adopting:
  - Inventory management,
  - Production-process modification,
  - Volume reduction,
  - Recovery and reuse
  - Sustainable product design

Inventory Management

- Proper control over the materials used in the manufacturing process is an important way to reduce waste generation. By reducing both the quantity of hazardous materials used in the process and the amount of excess raw materials in stock, the quantity of waste generated can be reduced. This can be done in two ways i.e. establishing material-purchase review and control procedures and inventory tracking system.
- Developing review procedures for all material purchased is the first step in establishing an inventory management program. Procedures should require that all materials be approved prior to purchase. In the approval process all production materials are evaluated to examine if they contain hazardous constituents and whether alternative non-hazardous materials are available.
- Another inventory management procedure for waste reduction is to ensure that only the needed quantity of a material is ordered. This will require the establishment of a strict inventory tracking system. Purchase procedures must be implemented which ensure that materials are ordered only on an as-needed basis and that only the amount needed for a specific period of time is ordered.

Production-Process Modification

- Changes can be made in the production process, which will reduce waste generation. This reduction can be accomplished by changing the materials used to make the product or by the more efficient use of input materials in the production process or both.
- Potential waste minimization techniques can be broken down into three categories:
  (i) Improved operating and maintenance procedures,
  (ii) Material change and
  (iii) Process-equipment modification.
- Improvements in the operation and maintenance of process equipment can result in significant waste reduction. This can be accomplished by reviewing current operational procedures or lack of procedures and examination of the production process for ways to improve its efficiency. Instituting standard operation procedures can optimise the use of raw materials in the production process and reduce the potential for materials to be lost through leaks and spills.

- A strict maintenance program, which stresses corrective maintenance, can reduce waste generation caused by equipment failure. An employee-training program is a key element of any waste reduction program. Training should include correct operating and handling procedures, proper equipment use, recommended maintenance and inspection schedules, correct process control specifications and proper management of waste materials.
- Hazardous materials used in either a product formulation or a production process may be replaced with a less hazardous or non-hazardous material. This is a very widely used technique and is applicable to most manufacturing processes. Implementation of this waste reduction technique may require only some minor process adjustments or it may require extensive new process equipment. For example, a circuit board manufacturer can replace solvent-based product with water-based flux and simultaneously replace solvent vapour degreaser with detergent parts washer.
- Installing more efficient process equipment or modifying existing equipment to take advantage of better production techniques can significantly reduce waste generation. New or updated equipment can use process materials more efficiently producing less waste. Additionally, such efficiency reduces the number of rejected or off-specification products, thereby reducing the amount of material which has to be reworked or disposed off. Modifying existing process equipment can be a very cost-effective method of reducing waste generation. In many cases, the modification can just be, relatively simple changes in the way the materials are handled within the process to ensure that they are not wasted.
- For example, in many electronic manufacturing operations, which involve coating a product, such as electroplating or painting, chemicals are used to strip off coating from rejected products so that they can be recoated. These chemicals, which can include acids, caustics, cyanides etc. are often a hazardous waste and must be properly managed. By reducing the number of parts that have to be reworked, the quantity of waste can be significantly reduced.

Volume Reduction

- Volume reduction includes those techniques that remove the hazardous portion of a waste from a non-hazardous portion. These techniques are usually to reduce the volume, and thus the cost of disposing off a waste material.
- The techniques that can be used to reduce waste-stream volume can be divided into 2 general categories: source segregation and waste concentration.
- Segregation of wastes is in many cases a simple and economical technique for waste reduction. Wastes containing different types of metals can be treated separately so that the metal value in the sludge can be recovered. Concentration of a waste stream may increase the likelihood that the material can be recycled or reused. Methods include gravity and vacuum filtration, ultra filtration, reverse osmosis, freeze vapourization etc.
- For example, an electronic component manufacturer can use compaction equipments to reduce volume of waste cathode ray-tube.

Recovery and Reuse

- This technique could eliminate waste disposal costs, reduce raw material costs and provide income from a salable waste. Waste can be recovered on-site, or at an off-site recovery facility, or through inter-industry exchange.
- A number of physical and chemical techniques are available to reclaim a waste material such as reverse osmosis, electrolysis, condensation, electrolytic recovery, filtration, centrifugation etc. For example, a printed-circuit board manufacturer can use electrolytic recovery to reclaim metals from copper and tin-lead plating bath.
- However, recycling of hazardous products has little environmental benefit if it simply moves the hazards into secondary products that eventually have to be disposed off. Unless the goal is, to redesign the product to use non-hazardous materials, such recycling is a false solution.

Sustainable Product Design

- Minimization of hazardous wastes should be at product design stage itself keeping in mind the following factors:
  - **(a) Rethink the product design:** Efforts should be made to design a product with fewer amounts of hazardous materials. For example, the efforts to reduce material use are reflected in some new computer designs that are flatter, lighter and more integrated. Other companies propose centralized networks similar to the telephone system.
  - **(b) Use of renewable materials and energy:** Bio-based plastics are plastics made with plant-based chemicals or plant-produced polymers rather than from petrochemicals. Bio-based toners, glues and inks are used more frequently. Solar computers also exist but they are currently very expensive.
  - **(c) Use of non-renewable materials that are safer:** Because many of the materials used are non-renewable, designers could ensure the product is built for re-use, repair and/or upgradeability. Some computer manufacturers such as Dell and Gateway lease out their products thereby ensuring they get them back to further upgrade and lease out again.

Method for Recycling of E-waste: (S-17)

(1) Recycling of e-waste: Monitors and CRT, keyboards, laptops, modems, telephone bards, hard drives, floppy drives, compact disk, mobiles, fax machines, printers, CPUs, memory chips, connecting wires and cables can be recycled. Recycling involves dismantling and recovery of valuable materials. Recycling is the best possible option for the management of e-waste because the existing dumping grounds in India are full and overflowing beyond capacity and it is difficult to get new dumping sites due to scarcity of land.

(2) Re-use: It is commonly used for electronic equipments like computers, cell phones etc. It constitutes direct second hand use or use after slight modification to the original functioning equipment. This method also reduces the volume of e-waste generation.

5.8 DISPOSAL OF E-WASTE (S-16; W-16, 18)

Disposal of E-waste is done by the following four methods:

1. Land filling: It is one of the most widely used methods for disposal of e-waste. In land filling, trenches are made on the flat surfaces. Soil is excavated from the trenches and waste material is buried in it, which is covered by a thick layer of soil. Modern techniques like secure landfill are provided with some facilities like, impervious liner made up of plastic or clay, leachate collection basin that collects and transfer the leachate to wastewater treatment plant. The degradation processes in landfills are very complicated and run over a wide time span.

The environmental risks from land filling of e-waste cannot be neglected because the conditions in a landfill site are different from a native soil, particularly concerning the leaching behaviour of metals. Mercury, cadmium and lead are the most toxic leachate.

Lead has been found to leach from broken lead containing glass, such as the cone glass of cathode ray tubes from TVs and monitors. Cadmium also leaches into soil and ground water. In addition, it is known that cadmium and mercury are emitted in diffuse form or via the landfill gas combustion plant. Landfills are also prone to uncontrolled fires, which can release toxic fumes. Therefore, land filling does not appear to be an environmentally sound treatment method for substances, which are volatile and not biologically degradable (Cd, Hg), persistent (Poly Chlorinated Biphenyls) or with unknown behaviour in a landfill site (brominated flame retardants).

Fig. 5.1

2. Incineration: It is a controlled and complete combustion process, in which the waste material is burned in specially designed incinerators at a high temperature (900-1000°C).

Advantages of incineration of e-waste are the reduction of waste volume and the utilization of the energy content of combustible materials. Some plants remove iron from the slag for recycling. By incineration some environmentally hazardous organic substances are converted into less hazardous compounds.

Disadvantages of incineration are the emission to air of substances escaping flue gas cleaning and the large amount of residues from gas cleaning and combustion. e-waste incineration plants contribute significantly to the annual emissions of cadmium and mercury. In addition, heavy metals not emitted into the atmosphere are transferred to slag and exhaust gas residues and can re-enter the environment on disposal. Therefore, e-waste incineration will increase these emissions, if no reduction measures like removal of heavy metals are taken.

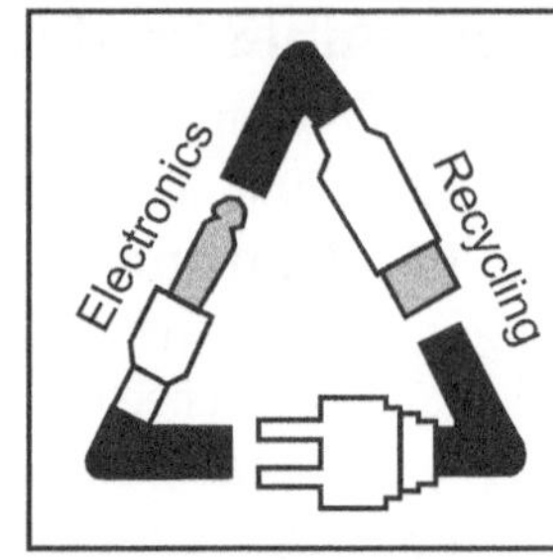

Fig. 5.2

3. Recycling of e-waste (W-18): Monitors and CRT, keyboards, laptops, modems, telephone boards, hard drives, floppy drives, compact disks, mobiles, fax machines, printers, CPUs, memory chips, connecting wires and cables can be recycled.

Recycling involves dismantling and recovery of valuable materials. i.e. removal of different parts of e-waste containing dangerous substances like PCB, Hg, separation of plastic, removal of CRT, segregation of ferrous and nonferrous metals and printed circuit boards.

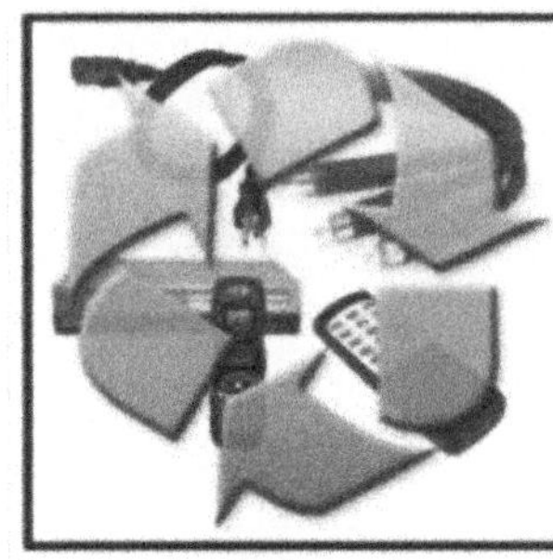
Fig. 5.3

Recyclers use strong acids to remove precious metals such as copper, lead, gold. The value of recycling from the element could be much higher if appropriate technologies are used.

The existing dumping grounds in India are full and overflowing beyond capacity and it is difficult to get new dumping sites due to scarcity of land. Therefore recycling is the best possible option for the management of e-waste.

4. Re-use: It constitutes direct second hand use or use after slight modifications to the original functioning equipment. It is commonly used for electronic equipments like computers, cell phones etc. Inkjet cartridge is also used after refilling. This method also reduces the volume of e-waste generation. We can use above mentioned methods for treatment and disposal of e-waste. The better option is to avoid its generation. To achieve this, buyback of old electronic equipments shall be made mandatory. This can considerably reduce the volume of e-waste generation.

Important Points

- Environmental pollution is the major problem associated with rapid industrialisation, urbanisation and rise in living standards of people. For developing countries, industrialisation was must and still this activity very much demands to build self-reliant and plays a major role also in uplifting nation's economy.
- Examples of industrial wastes are:
  - Chemical solvents
  - Paints
  - Sandpaper
  - Paper products
  - Industrial by-products
  - Metals
  - Radioactive wastes
  - Red mud
  - Coal ash
- Many wastes are mixtures of hazardous and non-hazardous wastes. Much of their contents may even be water. By segregating key toxic constituents, isolating liquid fraction, keeping hazardous streams away from non-hazardous wastes, generator can save substantial amounts of money on disposal or find new opportunities for recycling and reuse of wastes.
- A two-stage approach should be thought of for waste management, e.g., (a) prevention and (b) control of environmental pollution. Prevention aims at minimisation of industrial wastes at source, while the latter stresses on treatment and disposal of wastes.
- *Waste management at source* means the reduction of volume of raw materials and the waste, both, by some means.

- "Electronic waste" may be defined as 'discarded computers, office electronic equipment, entertainment electronic devices, mobile phones, television sets and refrigerators'. This definition includes used electronics and electrical items which are destined for reuse, resale, salvage, recycling or disposal.
- All over the world, the quantity of electrical and electronic waste generated each year, especially computers and televisions, has assumed alarming proportions. In 2006, the International Association of Electronics Recyclers (IAER) projected that 3 billion electronic and electrical appliances would become e-waste by 2010. That would be equivalent to an average e-waste generation rate of 400 million units a year till 2010.
- **E-Cycle Recovery will NOT:**
  - Place your equipment in landfill.
  - Refurbish, sell or give away your items.
  - Ship to where it may be processed under poor health and safety conditions.

Practice Questions

1. Describe 'Control measures of industrial waste'.
2. What are the disposal methods for industrial solid waste?
3. What are the varieties of industrial solid waste? Explain.
4. How is the management of E-waste done?
5. How is the recycling of E-waste done?
6. Describe the methods of disposal of E-waste.
7. Enumerate the effects of E-Waste constituent on health.
8. Describe the varieties of E-waste.
9. Write about the E-waste generation in India.
10. Define E-waste. What is the composition of E-waste? Explain.
11. Describe the biomedical waste management technologies/processes.
12. Define biomedical waste and write down the components of it.

MSBTE Questions and Answers (As Per 'G' Scheme)

Summer 2016

1. State the control measures of industrial waste. **(4M)**

Ans. Refer Section 5.3.

2. Define E-waste. State the dangers of E-waste. **(4M)**

Ans. Refer Sections 5.5 and 5.5.5.

3. State the benefits of recycling of industrial waste. **(4M)**

Ans. Refer Section 5.4.

4. Describe method of collection and disposal of E-waste. **(4M)**

Ans. Refer Sections 5.7 and 5.8.

5. Explain the varieties of E-waste. **(4M)**

Ans. Refer Section 5.5.4.

Winter 2016

1. Define E-Waste. State various types of E-Wastes. Explain the methods of disposal of E-waste. **(8M)**

Ans. Definition: Refer Sections 5.5.1 and 5.5.4.

Methods of Disposal of E-waste: Refer Section 5.8.

2. Describe method of collection and disposal of industrial waste. **(8M)**

Ans. Refer Section 5.2.

3. Explain Recycling of industrial waste. **(4M)**

Ans. Refer Section 5.4.

Summer 2017

1. State the varieties of E-Waste and why it is dangerous ? **(4M)**

Ans. Refer Sections 5.5.4 and 5.5.5.

2. Explain the methods of disposal and recycling of E-waste. **(4M)**

Ans. Refer Section 5.8.

3. Explain the various control measures of industrial waste. **(4M)**

Ans. Refer Section 5.3.

Winter 2017

1. Define E-waste. State varieties of E-waste. Explain effects of E-waste on human health and environment. **(4M)**

Ans. Refer Sections 5.5.1, 5.5.4 and Table 5.4.

2. Explain control measures taken for industrial waste. **(4M)**

Ans. Refer Section 5.3.

3. Explain the following: Industrial waste. **(2M)**

Ans. Refer Section 5.0.

4. Explain disposal and recycling of E-waste. **(8M)**

Ans. Refer Section 5.8.

Summer 2018

1. Define 'E-Wastes'. Enumerate the effects of E-Waste constituents on human health. **(4M)**

Ans. Refer Section 5.5.1, Table 5.4.

2. Explain about varieties of E-Waste stating its sources of generation. **(4M)**

Ans. Refer Section 5.5.4.

3. State the control measures to be adopted about industrial solid waste regarding its generation, collection and adverse effects. **(4M)**

Ans. Refer Section 5.3.

Winter 2018

1. Explain collection and disposal of industrial waste. **(4M)**

Ans. Refer Section 5.2.

2. Explain disposal of E-waste. **(4M)**

Ans. Refer Section 5.8.

3. Explain the recycling of industrial waste. **(4M)**

Ans. Refer Section 5.4.

4. State benefits of recycling of E-waste. **(4M)**

Ans. Refer Section 5.8

***

Chapter **6...**

Legal Aspects of Solid Waste Management

Weightage of Marks = 06, Teaching Hours = 04

Syllabus

Legal Aspects :

6.1 Legal Aspects - Present Scenario

6.2 Municipal Solid Waste Management Rules, 2016

6.3 Biomedical Waste Management Rules, 2016

6.4 E-Waste Management Rules, 2016

6.5 Construction and Demolition Waste Management Rules, 2016

6.6 Hazardous and Other Wastes Management Rules, 2016

6.7 Plastic Waste Management Rules, 2016

6.8 Role of Central Pollution Control Board and Maharashtra Pollution Control Board in Management of Solid Waste from Various Sources.

About this Chapter

At the end of this chapter, students will be able to:

- Suggest the existing legal framework to regulate the given type of solid waste with justification.
- Explain the relevant major provisions of Municipal Solid Waste Management Rules, 2016 for disposal of the given type of solid waste.
- Explain the different major provisions of Biomedical Waste Management Rules, 2016 for managing the given type of bio-medical waste.
- Compile the major features of Construction and demolition Waste Management Rules, 2016 for the disposal of the given type of waste.
- Explain the salient features of Hazardous and other wastes Management Rules, 2016 for the disposal of the given type of waste.
- Explain Role of CPCB (Central Pollution Control Board) and MPCB (Maharashtra Pollution Control Board) in managing the given type of solid waste.

6.1 LEGAL ASPECTS : PRESENT SCENARIO

- India holds the third rank in the world among the countries producing municipal solid waste (MSW) with 110000 tons of waste per day. As it lacks proper MSW management system and infrastructure, it has very high urban waste generation rates per capita. It has faced many catastrophic miseries brought by MSW and its poor quality disposal system.

- For example, when a plague broke out in Surat, in 1994, "uncollected solid waste blocking drains" was to blame as per the official inquiry on its causes. The notorious 2005 flood in Mumbai was also caused due to drains being blocked by plastic bags (its direct consequence being the introduction of Maharashtra Plastic Carry Bag Rules, 2006).
- Anyone can find many other incidents and occurrences which can be linked with the ineffectiveness of municipal corporations/civic bodies in MSW management. As far as waste management policy is concerned, India has been well behind as compared with other countries of the world.
- Until 2000, India did not even have any law with regard to MSW exclusively. Environmental legislations such as Water (Prevention and Control of Pollution) Act, 1974, Air (Prevention and Control of Pollution) Act, 1981 and Environment Protection Act, 1986 were brought in but the subject of MSW was neglected legally.
- Certain rules like Hazardous Wastes (Management and Handling) Rules, 1989 and Biomedical Waste (Management and handling) Rules, 1998 dealt with the subject only in a limited way. This glaring overlook was compounded by the cash-strapped status and general indifference of the civic bodies towards maintaining a functional MSW disposal system.
- In year 2000, Municipal Solid Wastes (Management and Handling) Rules, 2000 for the first time effected that dealt comprehensively with municipal solid waste, and which have been in place for the past 16 years until Solid Waste Management Rules (SWM), 2016 replaced these in year 2016.
- While the 2000 and 2016 rules deal with MSW generally, the following rules cover specific categories of waste that overlap with MSW:
  1. Hazardous Wastes (Management and Handling) Rules, 1989 (amended in 2000 and 2003) (superseded).
  2. Biomedical Waste (Management and handling) Rules, 1998 (superseded).
  3. Recycled Plastics Manufacture and Usage Rules, 1999.
  4. Batteries (Management and Handling) Rules, 2001 (amended in 2010).
  5. Hazardous Wastes (Management, Handling, and Transboundary Movement) Rules, 2008 (amended on 21st July 2009, 23rd September 2009, 30th March 2010 and 13th August 2010) (superseded).
  6. E-waste (Management and Handling) Rules, 2011 (superseded).
  7. Plastic Waste (Management and Handling) Rules, 2011 (superseded).
  8. E-Waste (Management) Rules, 2016.
  9. Hazardous and Other Wastes (Management and Transboundary Movement) Rules, 2016.
  10. Construction and Demolition Waste Management Rules, 2016.
  11. Bio-medical Waste Management Rules, 2016.
  12. Plastic Waste Management Rules, 2016.

Technical Guidelines Given by CPCB:

Following technical guidelines are given by CPCB which deal with various aspects of waste management:

- Guidelines on Implementing Liabilities for Environmental Damages due to Handling and Disposal of Hazardous Waste and Penalty.
- Guidelines for Common Hazardous Waste Incineration Criteria for Hazardous Waste Landfills Protocol for Performance Evaluation and Monitoring of the Common Hazardous Waste Treatment Storage and Disposal Facilities including Common Hazardous Waste Incinerators.
- Guidelines for Setting up of Operating Facility: Hazardous Waste Management.
- Guidelines for Proper Functioning and Upkeep of Disposal Sites.
- Guidelines for Environmental Sound Recycling of Hazardous Waste as per Schedule-V of Hazardous Waste (Management Handling and Transboundary Movement) Rules, 2008.
- Guidelines for the Selection of Site for Landfilling.

- Guidelines for Transportation of Hazardous Wastes.
- Guidelines for Storage of Incinerable Hazardous Wastes by the Operators of Common Hazardous Waste Treatment, Storage and Disposal Facilities and Captive HW Incinerators.
- Guidelines for Conducting Environmental Impact Assessment: Site Selection for Common Hazardous Waste Management Facility Manual for 'Sampling, Analysis, and Characterization of Hazardous Wastes'.

6.2 MUNICIPAL SOLID WASTE MANAGEMENT RULES, 2016

Fig. 6.1: Solid waste

- The Union Ministry of Environment, Forests and Climate Change (MoEF and CC) recently notified the new Solid Waste Management Rules (SWM), 2016. These will replace the Municipal Solid Wastes (Management and Handling) Rules, 2000, which have been in place for the past 16 years.
- These rules are the sixth category of waste management rules brought out by the ministry, as it has earlier notified plastic, e-waste, biomedical, hazardous and construction and demolition waste management rules.
- The rules have not pushed for decentralized management of waste but have encouraged centralized treatment such as waste to energy, the present state of which is not good in the country. Also, the informal sector has been considerably neglected in the new rules.
- According to Union Minister of State for Environment, Forests and Climate Change, 62 million tonnes of waste is generated annually in the country at present, out of which 5.6 million tonnes is plastic waste, 0.17 million tonnes is biomedical waste, hazardous waste generation is 7.90 million tonnes per annum and 15 lakh tonnes is e-waste and only about 75-80 per cent of the municipal waste gets collected and only 22-28 per cent of this waste is processed and treated.
- The new rules are now applicable beyond municipal areas and have included urban agglomerations, census towns, notified industrial townships, areas under the control of Indian Railways, airports, special economic zones, places of pilgrimage, religious and historical importance, and State and Central Government organizations in their ambit.

6.2.1 Salient Features of the New SWM Rules, 2016

(i) Segregation at Source:

- The new rules have mandated the source segregation of waste in order to channelize the waste to wealth by recovery, reuse and recycle. Waste generators would now have to segregate waste into three streams - Biodegradables, Dry (Plastic, Paper, metal, Wood etc.) and Domestic Hazardous waste (Diapers, Napkins, Mosquito repellants, Cleaning agents etc.) before handing it over to the collector.

- Institutional generators, market associations, event organizers and hotels and restaurants have been directly made responsible for segregation and sorting the waste and manage in partnership with local bodies.
- In case of an event, or gathering of more than 100 persons at any licensed/ unlicensed place, the organizer will have to ensure segregation of waste at source and handing over of segregated waste to waste collector or agency, as specified by the local body.
- All hotels and restaurants will also be required to segregate biodegradable waste and set up a system of collection to ensure that such food waste is utilized for composting / bio-methanation.
- The rules mandate that, all resident welfare and market associations and gated communities with an area of above 5,000 sq. m. will have to segregate waste at source into material like plastic, tin, glass, paper and others and hand over recyclable material either to authorized waste-pickers and recyclers or to the urban local body.

(ii) Collection and Disposal of Sanitary Waste:

- The manufacturers or brand owners of sanitary napkins are responsible for awareness for proper disposal of such waste by the generator and shall provide a pouch or wrapper for disposal of each napkin or diapers along with the packet of their sanitary products.

(iii) Collect Back Scheme for Packaging Waste:

- As per the rules, brand owners who sale or market their products in packaging material which are non-biodegradable, should put in place a system to collect back the packaging waste generated due to their production.

(iv) User Fees for Collection:

- The new rules have given power to the local bodies across India to decide the user fees. Municipal authorities will levy user fees for collection, disposal and processing from bulk generators. As per the rules, the generator will have to pay "User Fee" to the waste collector and a "Spot Fine" for littering and non-segregation, the quantum of which will be decided by the local bodies.
- Also, the new rules have mentioned about the integration of rag pickers, waste pickers and kabadiwalas from the informal sector to the formal sector by the state government.
- The rules also stipulate zero tolerance for throwing; burning, or burying the solid waste generated on streets, open public spaces outside the generator's premises, or in the drain, or water bodies.

(v) Waste Processing and Treatment:

- As per the new rules, it has been advised that the bio-degradable waste should be processed, treated and disposed off through composting or bio-methanation within the premises as far as possible and the residual waste shall be given to the waste collectors or agency as directed by the local authority.
- The developers of special economic zone, industrial estate, industrial park to earmark at least 5 per cent of the total area of the plot or minimum 5 plots/ sheds for recovery and recycling facility.
- Waste processing facilities will have to be set up by all local bodies having a population of 1 million or more within two years. For census towns with a population below 1 million or for all local bodies having a population of 0.5 million or more, common, or stand-alone sanitary landfills will have to be set up in three years time. Also, common, or regional sanitary landfills to be set up by all local bodies and census towns with a population under 0.5 million will have to be completed in three years.
- Also, the rules have mandated bio-remediation or capping of old and abandoned dump sites within five years.

(vi) Promoting Use of Compost:

- As per the rules, the Department of Fertilizers, Ministry of Chemicals and Fertilizers should provide market development assistance on city compost and ensure promotion of co-marketing of compost with chemical fertilizers in the ratio of 3-4 bags is to 6-7 bags by the fertilizer companies to the extent compost is made available for marketing to the companies.
- Also, the Ministry of Agriculture should provide flexibility in Fertilizer Control Order for manufacturing and sale of compost, propagating use of compost on farm land, set up laboratories to test quality of compost produced by local authorities or their authorized agencies.

(vii) Promotion of Waste to Energy:

- In a not-so welcoming move, the SWM Rules, 2016 emphasize promotion of waste to energy plants. The rules mandate all industrial units using fuel and located within 100 km from a solid waste-based Refuse-Derived Fuel (RDF) plant to make arrangements within six months from the date of notification of these rules to replace at least 5 per cent of their fuel requirement by RDF so produced.
- The rules also direct that non-recyclable waste having calorific value of 1500 kcal/kg or more shall be utilized for generating energy either through RDF not disposed off on landfills and can only be utilized for generating energy either or through refuse derived fuel or by giving away as feed stock for preparing refuse derived fuel. High calorific wastes shall be used for co-processing in cement or thermal power plants.
- As per the rules, the Ministry of New and Renewable Energy Sources should facilitate infrastructure creation for Waste to Energy plants and provide appropriate subsidy or incentives for such Waste to Energy plants.
- The Ministry of Power should fix tariff or charges for the power generated from the Waste to Energy plants based on solid waste and ensure compulsory purchase of power generated from such Waste to Energy plants by discoms.

(viii) Revision of Parameters and Existing Standards:

- As per the new rules, the landfill site shall be 100 metres away from a river, 200 metres from a pond, 200 metres away from highways, habitations, public parks and water supply wells and 20 km away from airports/airbase. Emission standards are completely amended and include parameters for dioxins, furans, reduced limits for particulate matters from 150 to 100 and now 50. Also, the compost standards have been amended to align with Fertilizer Control Order.

(ix) Management of Waste in Hilly Areas:

- As per the new rules, construction of landfills on hills shall be avoided. Land for construction of sanitary landfills in hilly areas will be identified in the plain areas, within 25 kilometers. However, transfer stations and processing facilities shall be operational in the hilly areas.

(x) Constitution of a Central Monitoring Committee:

- The government has also constituted a Central Monitoring Committee under the chairmanship of Secretary, MoEF and CC to monitor the overall implementation of the rules. The Committee comprising of various stakeholders from the Central and state governments will meet once a year to monitor the implementation of these rules.

Way Ahead:

- It will take almost 4-5 years to see the drastic change in how the waste management regimes will work in India. The SWM Rules, 2016 diminish hopes in pushing for adoption of a decentralized mechanism for solid waste management.
- However, it would be challenging to see how segregation at source shall work on the ground. A massive awareness campaign in association with communities, NGOs, students and other stakeholders needs to be planned to push for better implementation of these rules. The Rules need to focus on making solid waste management a people's movement by taking the issues, concerns and management of solid waste to citizens and grass-roots.

6.3 BIO-MEDICAL WASTE MANAGEMENT RULES, 2016

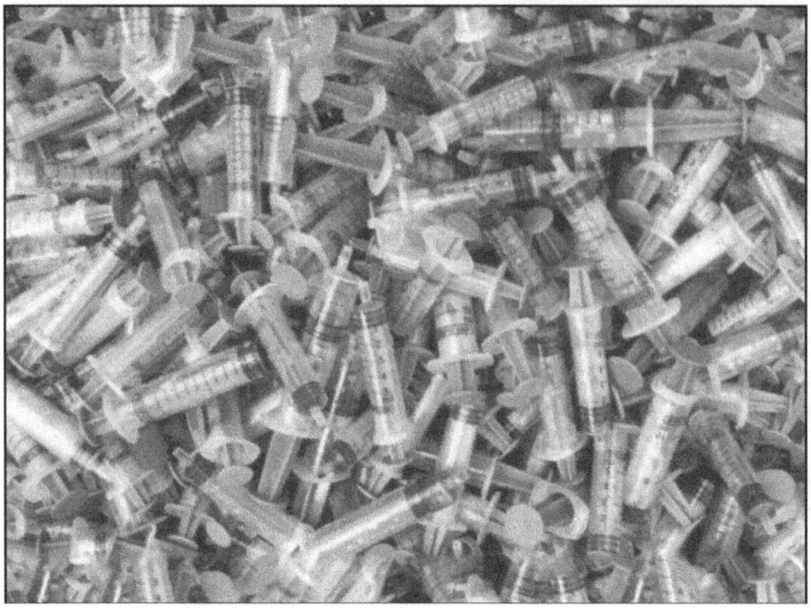

Fig. 6.2: Biomedical waste

- The environment ministry released the new Bio-medical Waste Management Rules, 2016 which bring in a wider and more comprehensive regime for bio waste management. The new bio-medical waste management rules change the way country used to manage this waste earlier.
- Under the new regime, the coverage has increased and also provides for pre-treatment of lab waste, blood samples etc. It mandates bar code system for proper control. It has simplified categorization and authorization. Thus, it makes a big difference to clean India Mission. This is a significant step forward, as the existing rules of 1998 have been revamped with major changes occurring in the reduction in waste categorization from the existing 10 categories to four.

6.3.1 Bio-medical Waste

- It is estimated that 484 tonnes of bio-medical waste are generated every day in India and out of this, 447 tonnes are treated. The rest of it pollutes the environment. As such kind of waste contains infectious pathogens, it should be ensured that it is treated properly.
- The new rules have made thing simple for hospitals as a reduction in the waste categorization number will lessen the confusion faced by waste handlers and other healthcare workers, who deal with bio-medical waste.
- The new rules seem to be comprehensive as these clearly define the roles of stakeholders, waste-generators and waste treatment facility operators, local authorities, state governments and pollution control boards.

Some Key Highlights:

- The new rules mandate a bar code system for bags and containers containing bio-medical waste which is a good move as this will restrict the entry of waste in the illegal recycling market. It will also be beneficial for rag-pickers and people dealing with infected solid plastic waste and working in recycling industries.
- The purview of the new rules has been expanded to include vaccination camps, blood donation camps, surgical camps and other healthcare activities.
- Under the new rules, bedded hospitals will get automatic authorization while there would be a one-time authorization for non-bedded ones.
- The new rules mandate pre-treatment of lab waste, blood samples, laboratory waste, microbiological waste and blood bags through disinfection or on-site sterilization as prescribed by WHO and NACO.
- Use of chlorinated plastic bags, gloves and blood bags will be phased out within two years. This will be a good step if implemented within the stipulated time of two years as burning of

chlorinated plastics often leads to emission of dioxin-furan, a persistent organic pollutant that has the potential to cause reproductive and developmental problems. It damages the immune system, interferes with hormones and causes cancer.

- The rules mandate training and regular immunization of all healthcare workers.

Scaling up Action:

- It is essential that implementation should follow the legal mandate. The NGOs visits to medical facilities have revealed shocking facts on the current handling and disposal practices of bio-medical waste which raise serious doubts on the government estimate of 92 per cent compliance level at present.
- Besides hospitals and clinics, bio-medical waste is generated in blood donation camps, vaccination camps and a host of other places. Therefore, the move of including these camps in the ambit of medical waste is a good decision. However, an action plan needs to be designed on how to monitor these camps and the waste they generate.
- The good thing is hospitals and Common Bio Medical Waste Treatment Facility (CBWTF) in Delhi and other cities across India have managed waste well.
- In India, there are around 198 CBWTFs while 28 new facilities are coming up. This ensures handling and managing bio-medical waste in a sustainable manner.

The Way Forward:

- As per the notification, all CBWTFs will have to ensure that operational standards are notified for running incinerators.
- Hospitals will need to phase out use of chlorinated plastic bags, gloves and blood bags within two years from the date of notification of these rules.
- Being a signatory to the Stockholm Convention, India may benefit from the phasing out of chlorinated plastic bags, gloves and blood bags, as most of the times these items end up in incinerators, causing huge emission of dioxin and furans.
- Hospitals have been asked to provide training to all their healthcare workers and others involved in the handling of bio-medical waste at the time of induction and after that at least once every year.
- They also have to report on the details of training programs conducted, the number of personnel trained and those who do not undergo training in the annual report.
- Hospitals have been directed to immunize all healthcare workers and others involved in the handling of bio-medical waste for protection against diseases.

6.4 E-WASTE (MANAGEMENT) RULES, 2016

Fig. 6.3: E-waste

- The Ministry of Environment, Forest and Climate Change has notified the E-Waste Management Rules, 2016 in supersession of the e-waste (Management and Handling) Rules, 2011. Announcing the notification of the rules, Minister of State (Independent Charge) of Environment, Forest and Climate Change, said that norms have been made more stringent and reflect the government's commitment to environmental governance. He pointed out that the E-waste rules will now include Compact Fluorescent Lamp (CFL) and other mercury containing lamps, as well as other such equipment.
- The Minister said that for the first time, the Rules will bring the producers under Extended Producer Responsibility (EPR), along with targets. He added that, producers have been made responsible for collection of E-waste and for its exchange.
- The bulk consumers must collect the items and handover them to authorized recyclers. He emphasized that various producers can have a separate Producer Responsibility Organization (PRO) and ensure collection of E-waste, as well as its disposal in an environmentally sound manner.
- The Minister pointed out that the role of State Governments has been also introduced to ensure safety, health and skill development of the workers involved in dismantling and recycling operations.

Salient Features:

- The E-wastes Management Rules, 2016 and as compared to the E-Waste (Management and Handling) Rules, 2011, there seem to be a significant change for good.
- The key difference between the previous set and the current one is the "reverse chain" that is mandated to be created by the producers under the Extended Producer Responsibility (EPR) to ensure electronic products are recovered for safe recycling.
- A "buy back" policy for electronics has also been suggested meaning that the producers can buy the products back for safe recycling by providing some sort of incentive to encourage consumers to also help recycle e-waste.
- Furthermore, the new set has taken cognizance of and has tried to include the informal sector which handles over 90 per cent of e-waste generated in India, in the mainstream.
- The informal sector will be formalised and the workers will be trained to handle e-waste rather than burning them after extracting the precious metals from them.
- This will fall under the state's responsibility to train the workers of the informal sector and including them in the mainstream so as to avoid leakages to the e-waste generated and recycled.
- The rules have been extended to the bulk consumers like companies that buy electronic goods in masses. The bulk consumers must collect the items and hand them over to authorized recyclers and it will be the responsibility of the one buying in bulk.
- The new rules have for the first time included the Compact Fluorescent Lamp (CFL) and other mercury containing lamps and equipment in their ambit. However, the disposal mechanism has not been clearly mentioned thus far but it will be done "scientifically".
- The draft notification that was released in June last year did not elaborate on the penal clauses. In fact, the penalty and punishment for non-compliance were the same as mentioned under the sections 15 and 16 of the Environment (Protection) Act, 1986. The sections say that an imprisonment term that may extend to five years or a fine which may extend to ₹ 1 lakh would be the penalty for non-compliance. The penalty clauses under the 2015 rules are not very clear.
- The 584 suggestions and objections were received to the draft rules and all the stakeholders including the industrial bodies were involved in drafting the final rules.
- India ranks fifth in the world in generating e-waste according to the UN's Global E-Waste Monitor. While the US generated 7.795 million tonnes in 2014, China generated a whopping 6.65 million tones, Japan stood third by generating 2.43 million tones. Germany ranked fourth with 1.95 million tonnes of e-waste and was closely followed by India generating 1.81 million tonnes. The Monitor adds that the waste generated by India is only slated to increase at an average of five per cent every year.

6.5 CONSTRUCTION AND DEMOLITION WASTE MANAGEMENT RULES, 2016

Fig. 6.4: Construction and demolition of waste

- The Government has notified Construction and Demolition Waste Management Rules, 2016 for the first time. Outlining the salient features of the Construction and Waste Management Rules, Union Minister of State (Independent Charge) of Environment, Forest and Climate Change, said that the rules are an initiative to effectively tackle the issues of pollution and waste management.
- The Minister said that at present, the construction and demolition waste generated is about 530 million tonnes annually. The Environment Minister said that, construction and demolition waste is not a waste, but a resource. He added that, the basis of these Rules is to recover, recycle and reuse the waste generated through construction and demolition.
- The Environment Minister said that, segregating construction and demolition waste and depositing it to the collection centers for processing will now be the responsibility of every waste generator. The Environment Minister highlighted that the local bodies will have to utilize 10-20% material from construction and demolition waste in municipal and government contracts.
- He said that, cities with a population of more than one million will commission processing and disposal facility within 18 months from the date of final notification of these rules, while cities with a population of 0.5 to 1 million and those with a population of less than 0.5 million will have to provide these facilities within two years and three years respectively.
- Permission for construction will be given only when the complete construction and demolition waste management plan is presented, he said. The Minister also pointed out that the large generators of waste will have to pay relevant charges for collection, transportation, processing and disposal, as notified by the concerned authorities.

Salient Features:

- Till now, rules for the management of construction and demolition waste were part of Schedule III of the rules for Municipal Solid Waste (Management and Handling), 2000. The ministry separately prepared draft rules for the management of C and D waste. The final notification of these draft rules takes into account nearly 111 suggestions from various environmentalists and environmental organizations.
- According to the rules, all generators of C and D waste must segregate it into four categories– concrete, soil, steel and wood, plastics, bricks and mortar – and then either deposit it at collection centres setup by the local authority or hand it over to processing facilities.

- But, out of all cities in the country, only Delhi has waste processing facilities – one in Burari and the other at Shahdara.
- Large generators (generating more than 20 tonnes of C and D waste a day or 300 tonnes of C and D waste per project in a month) must prepare an environment management plan detailing the environmental issues that can stem from the storage, transportation, disposal and reuse of C and D waste.
- They must also prepare a waste management plan, submit it to a local authority and get their approval before starting any construction/demolition/renovation work.
- They are also required to pay a relevant charge to service providers and contractors for collection and transport of C and D waste and to a waste processing facility for processing and disposal of this waste.
- Local authorities, in their turn, are required to use 10-20 per cent materials made from C and D waste in municipal and government contracts and give incentives to waste generators who process and recycle their waste *in-situ*.
- This is an important intervention that would need detailing with tax and pricing measures in cities. There are two ways of encouraging *in-situ* recycling. One is a 'waste tax' approach, in which you tax waste generators for availing waste processing facilities. The other is by incentivizing in-situ recycling.
- Further, the new rules mandate that all service providers and contractors are required to remove, and prepare a waste management plan for the waste generated in their jurisdiction within six months.
- Waste processing facilities, in their turn, are required to get an authorization from the State Pollution Control Board (SPCB) and notify a buffer zone of no development in the area surrounding their location if they have a capacity of processing more than five tonnes of C and D waste per day. Regardless of their capacity, it is advised that they are located away from habitation clusters, forest areas, water bodies, monuments, national parks and wetlands.
- For effective monitoring of stakeholders, the new rules also designate specific roles for the state government, the Central Pollution Control Board (CPCB), the Bureau of Indian Standards (BIS), the Indian Roads Congress (IRC) and central ministries.
- The concerned department of the state government is required to provide land for storage, processing and recycling of C and D waste within one and a half years. The CPCB must prepare operational guidelines on environmental management of C and D waste.
- The new rules also direct the IRC to prepare standards and practices on use of recycled C and D waste products in road construction.
- Central ministries such as the Ministry of Urban Development (MoUD), the Ministry of Rural Development (MoRD) and the Ministry of Panchayati Raj (MoPR) will facilitate local bodies while the Ministry of Environment and Forests (MoEF) will review implementation of these rules. The BIS should prepare a code of practice and standards for use of recycled C and D waste.
- Earlier, BIS did not recognize recycled C and D waste as suitable construction material. In January 2016, they notified an amendment in IS : 383 saying that recycled concrete can be used as a part substitute for concrete in constructions. BIS should now extend the scope of reuse of C and D waste to include its use as a part substitute for mortar and in the construction of paver blocks, kerb stones etc.
- The rules also set targets for cities to commission processing and disposal of C and D waste. These are one and a half years for million plus cities, two years for cities with a population between 0.5 million and one million and three years for cities with a population less than 0.5 million.
- Delhi recently issued an advisory on reuse of C and D waste, according to which, government agencies are required to incorporate a clause in their tenders, mandating that 10 per cent of recycled products from C and D waste should be used in road works, five per cent in non-structural applications and minimum 2 per cent in building works. It also sanctioned three new recycling plants in the state budget 2016-17. Other cities must follow suit.
- India generates nearly 530 million tonnes of C and D waste annually. This number is only going to skyrocket in the future, considering that two-thirds of the buildings that India will have in 2030 are yet to be built. Although inert, C and D waste is bulky and it usually finds its way into water bodies, public spaces and green areas. Recycled C and D waste can also be a great substitute for material mined from nature, thus reducing our dependence on fresh resources. Therefore, proper management of C and D waste is important.

6.6 HAZARDOUS AND OTHER WASTES (MANAGEMENT AND TRANS-BOUNDARY MOVEMENT) RULES, 2016

Fig. 6.5: Hazardous and other waste

- In order to strengthen the implementation of environmentally sound management of hazardous waste in the country, the Ministry of Environment, Forest and Climate Change has amended the Hazardous and Other Wastes (Management and Trans-boundary Movement) Rules, 2016.

Salient Features:

- Solid plastic waste has been prohibited from import into the country including in Special Economic Zones (SEZ) and by Export Oriented Units (EOU).
- Exporters of silk waste have now been given exemption from requiring permission from the Ministry of Environment, Forest and Climate Change.
- Electrical and electronic assemblies and components manufactured in and exported from India, if found defective can now be imported back into the country, within a year of export, without obtaining permission from the Ministry of Environment, Forest and Climate Change.
- Industries which do not require consent under Water (Prevention and Control of Pollution) Act 1974 and Air (Prevention and Control of Pollution) Act 1981, are now exempted from requiring authorization also under the Hazardous and Other Wastes (Management and Trans-boundary Movement) Rules, 2016, provided that hazardous and other wastes generated by such industries are handed over to the authorized actual users, waste collectors or disposal facilities.

Background:

- India's imports of solid plastic waste quadrupled (increased fourfold) from 12,000 tonnes in the fiscal year 2016-17 to 48,000 tonnes in FY 2017-18.
- India banned the import of plastic waste – particularly PET bottles – in 2015 because of an inability to recycle them.
- Experts suggest that most of the plastic waste does not make it to the recycling centres due to lack of an efficient waste segregation and inadequate collection.
- However, the Centre allowed agencies in special economic zones (SEZ) to import them in 2016. This was the legal loophole which led to an influx of PET bottles into the country.

6.6.1 PET Bottles

- PET, which stands for polyethylene terephthalate, is a form of polyester (just like the clothing fabric). It is extruded or molded into plastic bottles and containers for packaging foods and beverages etc.
- It is highly valued packaging material because it is strong yet lightweight, non-reactive, economical and shatterproof.
- Moreover, China, which was once a major global importer of plastic waste for recycling, had recently banned imports and therefore plastic wastes could be making their way to India instead.
- Plastic unsuitable for recycling is burnt, which releases toxic chemicals into the atmosphere. Or it ends up in landfill, potentially contaminating soil and water sources.

6.6.2 India's Plastic Waste Industry

- Before China banned the import of plastic waste, India was the 10th largest importer of plastic scrap. However, there was a "substantial increase" in import which shows that India has emerged as one of the alternatives for recycling plastic waste.
- India recycles a higher percentage of plastic waste internally than other richer countries.
- A 2012 study found that 56% of plastic waste produced in India is recycled.
- The process of collecting and separating plastic waste is largely an informal sector activity, providing employment to many.
- However, concerns remain regarding the widespread usage of plastic. Experts say, while recycling is a good practice, the focus must remain on reducing usage.
- While several states have indeed banned the usage of plastic bags, the regulations soon come undone.
- Maharashtra's ban on plastic items failed because of intense lobbying, lack of alternatives and also damage to the livelihood of those depending on manufacturing plastic and collecting plastic waste.

6.7 PLASTIC WASTE MANAGEMENT RULES, 2016

- The Government has notified the Plastic Waste Management Rules, 2016, in suppression of the earlier Plastic Waste (Management and Handling) Rules, 2011. The Minister of State for Environment, Forest and Climate Change, Shri Prakash Javadekar, said that the minimum thickness of plastic carry bags has been increased from 40 microns to 50 microns. He stated that 15, 000 tonnes of plastic waste is generated every day, out of which 9,000 tonnes is collected and processed, but 6,000 tonnes of plastic waste is not being collected.

Fig. 6.6: Plastic waste

- Shri Javadekar also said that the rules, which were admissible upto municipal areas, have now been extended to all villages. The Minister said that notifying the new Plastic Waste Management Rules is a part of the revamping of all Waste Management Rules. This will help in achieving the vision of our Prime Minister of Swachchh Bharat and cleanliness is the essence of health and tourism.

Salient Features:

- Minimum thickness of plastic carry bags has been increased from 40 microns to 50 microns.
- The rules which will be implemented across the country in six months are not clear about the fine amount to be imposed on plastic manufacturers or how the monitoring system would be carried out to pinpoint violators.
- The new plastic waste management rules are aimed at reducing 6,000 tonnes of uncollected plastic waste generated daily by targeting manufacturers and industries by using a new principle called the Extended Producers' Responsibility (EPR) Act.
- Under this Act, industries that use plastic will have to bear the onus and pay towards the collection of plastics through the distribution system that they use for retailing their products. Those failing to abide by the norms will lose the certification necessary to carry on with the operation. However, the modalities of implementation have not been specified as yet.
- Producers and importers, who will introduce plastics in the market in the form of carry bags, multi-layered plastic sachets or for packaging purpose need to establish a system for collecting back the waste generated as a result of their products.
- They will also have to work out the modalities for waste collection system based on the EPR through their own distribution channels or through the local bodies. The plan of collection has to be submitted to the respective state pollution control boards while applying for consent to establish, operate or renew the certification of business. This was done to improve the collection of plastic waste.

- One aspect that was not dealt with was the informal sector of waste collection. In Delhi, the informal sector employs about 150,000 people who transport almost 1,088 tonnes per day of recyclable waste.
- According to a survey carried out by Delhi-based non-profit Centre for Science and Environment, around 60–75 per cent of recyclable waste in present time consists of plastics. It used to be just 20–30 per cent four years ago.
- There is no mention on how to reduce plastic waste in the new rules. While it has focused on the use of plastic carry bags by increasing the minimum thickness from 40 microns to 50 microns, there has been no specification on the other forms of plastics such as the mineral water bottles (PET).

6.8 THE ROLE OF CPCB AND MPCB IN MANAGEMENT OF SOLID WASTE FROM VARIOUS SOURCES

- Since the disposal of municipal solid wastes poses problems of the pollution and health hazards, the Pollution Control Boards are expected to take action for persuading the civic authorities in proper management of municipal solid wastes.
- Though, direct responsibility of management of solid wastes is on the local municipal authorities, the Pollution Control Boards need to have close linkage with local authorities in rendering assistance in terms of carrying out necessary surveys and providing technological back-up.
- The Central Pollution Control Board and the Maharashtra Pollution Control Boards at the national and state levels are to disseminate information and create awareness among the concerned authorities and public at large.
- The Central Pollution Control Board (CPCB) and the Maharashtra Pollution Control Boards (MPCB) within the given powers to them under relevant Acts and Rules have been attempting to persuade local bodies to take appropriate measures for the treatment and disposal of municipal solid waste.
- In order to initiate a systematic approach on proper management of municipal waste (solid), CPCB issues directions to MPCB of relevant sections of various acts.
- The MPCB shall monitor the compliance of the standards regarding ground water, ambient air, leachate quality and the compost quality including incineration standards as specified under **Schedules II, III** and **IV**.
- The MPCB, after the receipt of application from the municipal authority or the operator of a facility in **Form I,** for grant of authorization for setting up waste processing and disposal facility including landfills, shall examine the proposal taking into consideration the views of other agencies like the State Urban Development Department, the Town and Country Planning Department, Air Port or Air Base Authority, the Ground Water Board or any such other agency prior to issuing the authorization.
- The MPCB shall issue the authorization in **Form-III** to the municipal authority or an operator of a facility within forty-five days stipulating compliance criteria and standards as specified in **Schedules II, III** and **IV** including such other conditions, as may be necessary.
- The authorization shall be valid for a given period and after the validity is over, a fresh authorization shall be required.

Important Points

- India holds the third rank in the world among the countries producing municipal solid waste (MSW) with 110000 tons of waste per day.
- In year 2000, Municipal Solid Wastes (Management and Handling) Rules, 2000 for the first time effected that dealt comprehensively with municipal solid waste, and which have been in place for the past 16 years until Solid Waste Management Rules (SWM), 2016 replaced these in year 2016.
- According to Union Minister of State for Environment, Forests and Climate Change, 62 million tonnes of waste is generated annually in the country at present, out of which 5.6 million tonnes is plastic waste, 0.17 million tonnes is biomedical waste, hazardous waste generation is 7.90 million tonnes per annum and 15 lakh tonnes is e-waste and only about 75-80 per cent of the municipal waste gets collected and only 22-28 per cent of this waste is processed and treated.

- Salient Features of the new SWM Rules, 2016 include segregation at source, collection and disposal of sanitary waste, collect back scheme for packaging waste, user fees for collection, waste processing and treatment, promoting use of compost, promotion of waste to energy, revision of parameters and existing standards, management of waste in hilly areas and constitution of a Central Monitoring Committee.
- The Rules need to focus on making solid waste management a people's movement by taking the issues, concerns and management of solid waste to citizens and grass-roots.
- It is estimated that 484 tonnes of bio-medical waste are generated every day in India and out of this, 447 tonnes are treated. The rest of it pollutes the environment. As such kind of waste contains infectious pathogens, it should be ensured that it is treated properly.
- The new rules have made thing simple for hospitals as a reduction in the waste categorization number will lessen the confusion faced by waste handlers and other healthcare workers, who deal with bio-medical waste.
- The new rules seem to be comprehensive as these clearly define the roles of stakeholders, waste-generators and waste treatment facility operators, local authorities, state governments and pollution control boards.
- The Ministry of Environment, Forest and Climate Change has notified the E-Waste Management Rules, 2016 in supersession of the e-waste (Management and Handling) Rules, 2011.
- The bulk consumers must collect the items and hand them over to authorized recyclers. He emphasized that various producers can have a separate Producer Responsibility Organization (PRO) and ensure collection of E-waste, as well as its disposal in an environmentally sound manner.
- The Government has notified Construction and Demolition Waste Management Rules, 2016 for the first time. The construction and demolition waste generated is about 530 million tonnes annually and construction and demolition waste is not a waste, but a resource. The basis of these Rules is to recover, recycle and reuse the waste generated through construction and demolition. Segregating construction and demolition waste and depositing it to the collection centres for processing will now be the responsibility of every waste generator. The local bodies will have to utilize 10-20% material from construction and demolition waste in municipal and government contracts.
- In order to strengthen the implementation of environmentally sound management of hazardous waste in the country, the Ministry of Environment, Forest and Climate Change has amended the Hazardous and Other Wastes (Management and Transboundary Movement) Rules, 2016.
- The Government has notified the Plastic Waste Management Rules, 2016, in suppression of the earlier Plastic Waste (Management and Handling) Rules, 2011. The minimum thickness of plastic carry bags has been increased from 40 microns to 50 microns. 15,000 tonnes of plastic waste is generated every day, out of which 9,000 tonnes is collected and processed, but 6,000 tonnes of plastic waste is not being collected. Plastic Waste Management Rules is a part of the revamping of all Waste Management Rules. This will help in achieving the vision of our Prime Minister of Swachchh Bharat.
- Since the disposal of municipal solid wastes poses problems of the pollution and health hazards, the Pollution Control Boards are expected to take action for persuading the civic authorities in proper management of municipal solid wastes.
- Though, direct responsibility of management of solid wastes is on the local municipal authorities, the Pollution Control Boards need to have close linkage with local authorities in rendering assistance in terms of carrying out necessary surveys and providing technological back-up.

Questions For Practice

1. State the salient features of Solid Waste Management Rules (SWM), 2016.
2. State the salient features of E-Waste (Management) Rules, 2016.
3. State the salient features of Hazardous and Other Wastes (Management and Trans-boundary Movement) Rules, 2016.
4. State the salient features of Construction and Demolition Waste Management Rules, 2016.
5. State the salient features of Bio-medical Waste Management Rules, 2016.
6. State the salient features of Plastic Waste Management Rules, 2016.
7. Describe the present scenario with regard to SWM in India.
8. State the Role of CPCB and MPCB in Management of Solid Waste from Various Sources.

***